Adeolu Aderoju
Oyelowo G. Bayowa
Olatunbosun A. Alao

Hydrocarbon Exploration Using Unconventional Interpretation Techniques

Reflection Seismology

Anchor Academic Publishing

Aderoju, Adeolu, Bayowa, Oyelowo G., Alao, Olatunbosun A.: Hydrocarbon Exploration
Using Unconventional Interpretation Techniques: Reflection Seismology, Hamburg,
Anchor Academic Publishing 2015

Buch-ISBN: 978-3-95489-378-2
PDF-eBook-ISBN: 978-3-95489-872-5
Druck/Herstellung: Anchor Academic Publishing, Hamburg, 2015

Bibliografische Information der Deutschen Nationalbibliothek:
Die Deutsche Nationalbibliothek verzeichnet diese Publikation in der Deutschen
Nationalbibliografie; detaillierte bibliografische Daten sind im Internet über
http://dnb.d-nb.de abrufbar.

Bibliographical Information of the German National Library:
The German National Library lists this publication in the German National Bibliography.
Detailed bibliographic data can be found at: http://dnb.d-nb.de

All rights reserved. This publication may not be reproduced, stored in a retrieval system
or transmitted, in any form or by any means, electronic, mechanical, photocopying,
recording or otherwise, without the prior permission of the publishers.

Das Werk einschließlich aller seiner Teile ist urheberrechtlich geschützt. Jede Verwertung
außerhalb der Grenzen des Urheberrechtsgesetzes ist ohne Zustimmung des Verlages
unzulässig und strafbar. Dies gilt insbesondere für Vervielfältigungen, Übersetzungen,
Mikroverfilmungen und die Einspeicherung und Bearbeitung in elektronischen Systemen.

Die Wiedergabe von Gebrauchsnamen, Handelsnamen, Warenbezeichnungen usw. in
diesem Werk berechtigt auch ohne besondere Kennzeichnung nicht zu der Annahme,
dass solche Namen im Sinne der Warenzeichen- und Markenschutz-Gesetzgebung als frei
zu betrachten wären und daher von jedermann benutzt werden dürften.

Die Informationen in diesem Werk wurden mit Sorgfalt erarbeitet. Dennoch können
Fehler nicht vollständig ausgeschlossen werden und die Diplomica Verlag GmbH, die
Autoren oder Übersetzer übernehmen keine juristische Verantwortung oder irgendeine
Haftung für evtl. verbliebene fehlerhafte Angaben und deren Folgen.

Alle Rechte vorbehalten

© Anchor Academic Publishing, Imprint der Diplomica Verlag GmbH
Hermannstal 119k, 22119 Hamburg
http://www.diplomica-verlag.de, Hamburg 2015
Printed in Germany

DEDICATION

This beautiful piece is strictly dedicated to my heavenly father; the owner of the universe, the author of my life, the one who knows my destination.

ACKNOWLEDGEMENT

My greatest acknowledgement goes to Lord Almighty, the giver of knowledge, wisdom, talent, inspiration, ability, grace and ideas. Also, I acknowledge my good parents and siblings, their support are unending. Your scribes are so bold and deep, it is indelible. I appreciate you all.

My profound appreciation also goes to my Mentors; Dr. O. G. Bayowa, Dr. O. A. Alao and Dr. A. A. Adepelumi. Their impacts, love, trust and the hope they have awaken in me will forever remain. I am grateful.

My fat appreciation goes to the Earth Sciences Department, Ladoke Akintola University of Technology, Ogbomoso, the members of Staff and the students alike for a wonderful time shared together. I love you all.

Also, I want to acknowledge the Geology Department of Obafemi Awolowo University, Ile – Ife; from which I retrieved the Data that I worked with. Thank you so much, I am grateful.

I am also indebted to a great number of my friends; who have shown their support in diverse ways during the course of my stay in Ladoke Akintola University of Technology, Ogbomos, Nigeria; my appreciation is unending – Temitayo Alabi, Olaniyi Ayobami, Wole Olufunmilayo, Seun Olaniyi, Timothy Afonja, Wole Eesuola, Wale Egbuna, and many more that have escaped this list.

TABLE OF CONTENT

DEDICATION 5
ACKNOWLEDGEMENT 6
ABSTRACT 12

CHAPTER 1 13
    1.1   Introduction 13
    1.2   Aim of Research 14
    1.3   Specific Objectives of the Study 14
    1.4   Methodology 15
    1.5   LOCATION AND DESCRIPTION OF THE STUDY AREA 16
    1.6   LITERATURE REVIEW 19

CHAPTER 2 22
    2.1   GEOLOGY OF THE NIGER DELTA 22
    2.2   DEVELOPMENT OF THE DELTA 25
    2.3   THE PRESENT NIGER DELTA 26
    2.4   SUBSURFACE FORMATIONS OF THE TETIARY NIGER DELTA 26
        2.4.1   Akata Formation 26
        2.4.2   Agbada Formation 28
        2.4.3   Benin Formation 29
    2.5   BASIN ARCHITECTURE 30
    2.6   SUBSURFACE STRUCTURES OF THE NIGER DELTA 33
        2.6.1   GROWTH FAULT 34
        2.6.2   ROLLOVER ANTICLINES 34
        2.6.3   TRAPS AND SEALS 36
    2.7   TECTONICS OF THE NIGER DELTA 37

CHAPTER 3 39
    3.1   MATERIALS AND METHOD OF STUDY 39
    3.2   PHYSICAL BASIS OF REFLECTION SEISMOLOGY 39
        3.2.1   P AND S WAVES 39
        3.2.3   REFLECTION AND PHYSICAL PROPERTIES OF ROCKS 41
        3.2.4   SEISMIC ACQUISITION 43

3.2.5	SEISMIC DATA PROCESSING	44
3.2.6	SEISMIC DATA INTERPRETATION	44
3.3	WELL LOGGING	48
3.3.1	BOREHOLE ENVIRONMENT AND INVASION PROFILES	50
3.3.2	LOG TYPES AND USES	50
3.3.3	WELL LOG CORRELATION	57
3.4	SEISMIC ATTRIBUTES	57
3.4.1	Introduction	57
3.4.2	The Classification of Attributes	58
3.5	Spectral Decomposition: A Powerful Risk Reduction Tool	61

CHAPTER 4		66
4.1	DISCUSSION AND RESULTS PRESENTATION.	66
4.2	Conventional Mapping	66
4.2.1	FAULT MAPPING	67
4.2.2	Horizon Tracking.	69
4.3	Well to Seismic tying	71
4.4	WELL LOG INTERPRETATION	76
4.4.1	Thin Sandstone Bed 1 (TB 01T):	77
4.4.2	Thin Sandstone bed 2 (TB 03T):	82
4.5	SPECTRAL DECOMPOSITION	87
4.5.1	Spectral Decomposition Analysis of TB 01T	89
4.5.2	Trapping Mechanism in TB 01T	101
4.5.3	Spectral Decomposition Analysis of TB 03T	101
4.5.4	Hydrocarbon Potential in TB 03T	105
4.6	ISO - CHRON MAPS	106
4.6.1	TB 01T ISOCHRON MAP: Time values to this horizon range 2437.4 ms to 2892.1 ms	107
4.6.2	TB 03T ISOCHRON MAP: Time values to this horizon range 2163.3 ms to 2457.3 ms.	107

CHAPTER 5		109
5.1	CONCLUSION	109
5. 2	RECCOMENDATION	109

REFERENCES	111

LIST OF TABLES

Table 1 The Properties of Sandstone bed TB 01T in relation to the wells it cuts across 80
Table 2 The Properties of Sandstone bed TB 03T in relation to the wells it cuts across 83

LIST OF FIGURES

Figure 1.1: Schematic Diagram Showing the Research Methodology... 16
Figure 1.2: Basemap of the Study Area; Inset is the Map of Nigeria... 17
Figure 1.3: Description of the Study Area. 18
Figure 2.1: Index Map of Nigeria and Cameroon 24
Figure 2.2: Stratigraphic columns showing the three formations of the Niger Delta.. 27
Figure 2.3: Stratigraphic Column Showing the Three Formation of the Niger Delta. 32
Figure 3.1: Simple Schematic Diagram Illustrating the Underlying Principles of the Seismic Method. 41
Figure 3.2: Definition of a Wavelet... 42
Figure 3.3: Reflected and Transmitted Rays Associated With a Ray Normally Incident on an Interface
of Acoustic Impedance Contrast.. 42
Figure 3.4: Simplified Processing Flow (After Yilmaz 1987) 46
Figure 3.5: Borehole Environments and Invasion Profiles (www.petrolog.net/webhelp/logging_tools)............ 51
Figure 3.6: Spectral decomposition is used to identify thin beds through analysis of the frequency spectrum
in a short window around the time of the bed. (Partyka *et al.*, 1999).. 63
Figure 3.7: Comparison of Frequency Effects on thin Beds... 64
Figure 3.8: Spectral Decomposition Work Flow. 65
Figure 4.1: Some of the picked faults on inline 5915 with Mapped pay sand reflectors (TB 01T and TB 3T)...... 68
Figure 4.3: Two tracked reflectors across the two markers; TB 03T and TB 01B.. 70
Figure 4.4: Reflector Horizon for TB 03T tracked across the volume with the used wells................................. 70
Figure 4.5: Reflector TB 01T tracked across the volume with the used wells.. 71
Figure 4.6: The Relationship between Sonic Log (DT), Computed Acoustic Impedance in Relation to
Depth and the Generated Synthetic Wave Form and Seismic Wave Form....................................... 72
Figure 4.7: Matching of the Synthetic Wave Form to the Seismic Wave Form... 73
Figure 4.8: An Excel Spread sheet, Showing the Generated Data for one of the Four Wells used for
tying and the Consequent Depth - Time Plot. 73
Figure 4.9: The Depth – Time Graph of Well 3. 74
Figure 4.10: The Depth – Time Graph of Well 4. 74
Figure 4.11: The Depth – Time Graph of Well 5. 75
Figure 4.12: The Depth – Time Graph of Well 6. 76
Figure 4.13: Showing the Cross Section of TB 01T across the wells. 78
Figure 4.14: Showing the cross section of TB 01T across the wells as it misses out in well 02. 78
Figure 4.15: Using a Better Resolution Scale to View TB 01T.. 80
Figure 4.16: TB 01T Sandwiched within a Thick Shale Unit 81
Figure 4.17: Showing the cross section of TB 03T across the wells.. 84
Figure 4.18: An improve scale enhance the viewing and mapping of the TB 03T. 85
Figure 4.19: TB 03T Sandwiched within a Thick Shale Unit 85
Figure 4.20: Seismic Data Showing the position of the Markers on Inline 5915- 88
Figure 4.21: Amplitude Attribute- Inline 5915. 88
Figure 4.22: Inline 5915- Energy Attribute 89
Figure 4.23: Inline 5915- 25Hz FFT 90

Figure 4.24: Inline 5915- 30Hz FFT 91
Figure 4.25: Inline 5915- 33Hz FFT 91
Figure 4.26: Inline 5915- Color blended Image; Blue- 25Hz, Green- 30 Hz, and Red- 33Hz...................................93
Figure 4.27: Inline 5915- Color blended of Amplitude Attribute, Blue- 25Hz, Green- 30 Hz, and Red- 33Hz.......93
Figure 4.28: Inline 5915- Color blended of Energy Attribute, Blue- 25Hz, Green- 30 Hz, and Red- 33Hz...........94
Figure 4.29: The seismic data of TB 01T 94
Figure 4.30: The 15Hz FFT of TB 01T 96
Figure 4.31: The 25Hz FFT of TB 01T 96
Figure 4.32: Showing the 30Hz FFT of TB 01T 97
Figure 4.33: Showing the 45Hz FFT of TB 01T 97
Figure 4.34: Amplitude Attribute Map of TB 01T 98
Figure 4.35: Energy Attribute Map of TB 01T 98
Figure 4.36: Color Blended Map of TB 01T - 25Hz/Blue, 30Hz/Green, 33Hz/Red...99
Figure 4.37: Color Blended Map of TB 01T - 25Hz/Blue, 30Hz/Green, 33Hz/Red, with Energy. 99
Figure 4.38: Color Blended Map of TB 01T - 25Hz/Blue, 30Hz/Green, 33Hz/Red, with Amplitude...................100
Figure 4.39: Inline 5915 Instantaneous Amplitude Attribute, overlain on it is well 5 and its log and
    Marker for TB 03T 102
Figure 4.40: Showing Inline 5915 for 25Hz, overlain on it is well 5 and its log and Marker for TB 03T.............103
Figure 4 .41: Showing Inline 5915 for 30Hz, overlain on it is well 5 and its log and Marker for TB 03T.............103
Figure 4.42: Instantaneous Amplitude Attribute on TB 03T. 104
Figure 4.43: 25Hz Frequency display on TB 03T...105
Figure 4.45: 33Hz Frequency display on TB 03T...106
Figure 4.46: Iso – Chron Map for TB 01T 107
Figure 4.47: Iso-Chron Map for TB 03T 108

ABSTRACT

Technology has been predicted as the last card in combating the ever increasing demand of our commodity; oil and gas. Modern approach to Seismic acquisitions and interpretation is of prime essence. A powerful

Interpretation tool; Spectral decomposition, has been used to visualize (in 3 – D) a thin sandstone reservoir in "X – Field" Niger Delta, Nigeria. The study aimed at mapping thin sandstone beds which have been considered sub seismic or un-mapable.

Well log data were acquired and qualitatively and quantitatively interpreted using RocDok software and correlated across four wells; Tmb 01, 02, 04 and 05. The interpreted well logs were presented in Tables. Seismic data were also acquired. Thin sandstone bed markers were identified from well logs and consequently mapped on the seismic data. The data were interpreted using the OpendTect software for frequency analysis which was presented in form of maps.

Two thin sandstone beds were delineated from the well log these are; TB 01T and TB 03T. TB 01T had an average thickness of 4.16m and cuts across three wells (Tmb 01, 04 and 05) but was not delineated in Tmb 02. TB 03T was extremely thin. Its average thickness could not be estimated but cut across the four wells. Frequency analysis from seismic sectional view revealed the portion of the thin pay sandstone as the markers were engulfed by high amplitude. The plan view seismic analysis also showed high amplitude across the region of the thin sandstone.

The study concluded that the high amplitude represented on the seismic views characterize the thin sandstone pay. However, other regions on the seismic views with high amplitude could be new prospects.

CHAPTER 1

1.1 Introduction

Over time, beds of relative thicknesses between 10 - 30m or thin reservoirs have been deemed unmapable or probable to interpreters using conventional interpretation techniques. This is caused by various factors, such as: the reservoir's thinness, discontinuous occurrences, high degree of vertical and lateral variability in net sand thickness, weak impedance contrast at sand interfaces, high impedance of bounding overlying layers and the limited bandwidth of seismic data.

In geological settings where there are successive intercalations of rocks such as the Niger Delta basin of Nigeria, mapping of such thin reservoir may seem more challenging using the conventional Technology. This project is set to map such thin pay zones using a 3-D visualization and spectral decomposition attributes which in turn will help to define the thickness, heterogeneity, rock property changes, depth thickness and the lateral spread across oil wells in the selected locale of the Niger Delta, Nigeria.

The Research problem

The term "All the easy oil has been found" is no more a novel term, it has grown popular. In the 1900s it was not uncommon to search for oil tucked beneath structurally simple anitclinal traps. However, geoscientists have to work much harder today to discover economic reservoirs. Searching for hard-to-find stratigraphic traps is common as we reexamine mature plays to uncover missed hydrocarbons. In modern times, simply relying on the full-stack amplitude response as a direct hydrocarbon indicator is not enough; typically, advanced reservoir characterization techniques and seismic attribute analyses are needed to evaluate a reservoir properly.

Most Seismic datasets contain higher frequencies near the surface; all frequencies attenuate as the wave front propagates deeper into the subsurface. Many of the Pays lie in depths where they are primarily imaged by frequencies as low as 10 to 20 Hz. This means that thin beds, beds that are 16 to 82 ft (5 to 25 m) thick are sub-seismic in nature (i.e., they cannot be resolved by the seismic wavelet and are considered to be below the tuning thickness). These are the type of thin stringer sands that are encountered on usual basis in seismic datasets. Given the low frequency content of these datasets, advanced attribute analyses techniques are required to properly evaluate these deposits. The processes helped to supplement the thin sandstone beds exploitation and to uncover the data that traditionally goes undetected.

1.2   Aim of Research

To map thin pay zones using 3-D visualization and spectral decomposition attribute.

1.3   Specific Objectives of the Study

The specific objectives of the study are to

1. Carry out interpretation of well data to determine possible thin reservoir sand, including the sand Reservoir in the study area.
2. Carry out conventional seismic interpretation over the study area, mapping faults and interpreting horizons
3. Overlay the well log data interpretation on the seismic interpretation to determine a lateral continuity of the thin sand.

4. Apply spectral decomposition to the interpreted seismic sections from 3 above and viewing the results in 3-D. thereby defining heterogeneity, rock property changes, depth thickness and the lateral spread across oil wells in the selected locale of study.
5. Identify and differentiate between reservoir and non reservoir Facies by a comprehensive integrated analysis of the spectral decomposition attributes.

1.4 Methodology

3-D seismic data was acquired in suit of Borehole log which is inclusive of Gamma-ray, Resistivity and Porosity logs. The 3-D seismic data and well log data were interpreted on RokDoc Software and OpendTect Software. A detailed interpretation of the well data was carried out using RokDoc Software to delineate possible Hydrocarbon bearing sands. The results of this well log interpretation were overlaid on interpreted seismic sections so as to delineate the lateral extent of the reservoir sands (including thin sands).

The composite interpretation of the Seismic interpretation on RokDoc Software was then exported to OpendTect Software. Spectral Decomposition attribute would then be applied to the exported interpretation on OpendTect software. The spectral decomposition Technique was used to decompose the seismic data with normal frequency bandwidth into a set of sections having discrete or very narrow bandwidth. The instantaneous frequency volume was produced in respect to the pay zone reflector. This was analyzed by a 3-D visualization tools. The frequency was extracted along the pay zone reflector and scanned so as to discover meaningful patterns by restricting the frequency.

The illumination of selected frequency band along the pay horizon was used to produce the "geologic appearance" and was consequently be used to prove the wells that are outside the frequency anomaly (i. e. wells that does not contain the pay sands, due to the pay's limited lateral spread). Finally, a 3-D visualization was done to differentiate between reservoir and non reservoir Facies (See the Schematic diagram of the Research Methodology in Figure 1.1).

1.5 LOCATION AND DESCRIPTION OF THE STUDY AREA

The study field, for the purpose of this study is identified as Tomboy field and is located between latitudes 0° 36' 0.3" to 0° 38' 59.4" North of the equator and longitudes 2° 47' 52.6" to 2° 53' 16.1" East of Greenwich Mean Time (GMT) (Figure 1.2).

The study area can be described in terms of the tectonic setting, and it is situated at the intersection of the Benue trough and the South Atlantic Ocean where a triple junction (Figure 1.3) developed during the separation of South America and Africa continents during the Mesozoic (Burke et al., 1972; Whiteman, 1982).

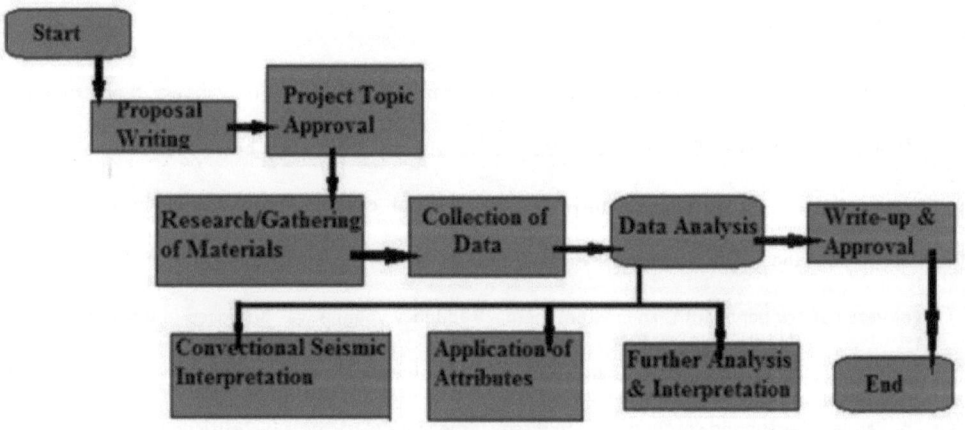

Figure 1.1: Schematic Diagram Showing the Research Methodology.

Figure 1.2: Basemap of the Study Area; Inset is the Map of Nigeria

Figure 1.3: Description of the Study Area.

1.6 LITERATURE REVIEW

Ekweozor and Okoye (1980) evaluated the petroleum source bed of the Niger Delta, supporting the conclusion of Weber and Daukoru (1975) that the source rocks are shale of the Akata Formation.

Orife and Avbovbo (1982) studied seismic sections of the Niger Delta and observed that hydrocarbons are trapped in stratigraphic traps such as crested accumulation below mature erosion surfaces, canyon-fill accumulations above unconformity surfaces, and traps due to facies change. These are in addition to the known structural traps of the roll-over anticlines and fault closures.

Ejedawe and Coker (1984) examined the evolution of oil generative window and occurrence of oil in the Niger Delta. They concluded that during the active subsidence phase, oil was generated initially at a temperature of $284 - 2950^0F$ and a depth of $9,840 - 17,060$ ft $(3,000 - 5,201$ m). However, after the subsidence there was vertical upward movement of the oil generative window through $2,625 - 5,250$ ft $(800 - 1,600$ m) accompanied by a temperature change of $41 - 910^0F$. This caused the maturation of the source rocks at progressively shallower depths and lower temperatures.

Nwachukwu and Chukwura, (1986) discussed intensely the organic matter of the Agbada Formation. They concluded that the shales of the Agbada Formation are mature and contains essentially Type III organic matter, which is capable of generating oil and gas. The average threshold temperature for the onset of oil generation was found to be 2150^0F. The over-pressured zones are characterized by porosity values being higher than expected, and the bulk density being correspondingly lower. This information is important when drilling for oil in over-pressured zones because they may cause blowouts if not properly controlled.

Ekweozor et al., (2000) studied the gas reserves of the Niger Delta and classified them into Associated Gas (AG) and Non-associated (NAG). They observed from thermal maturity values that these gases were formed contemporaneously with the produced oil in the Niger Delta. These gases also contain "Gas condensate" which are liquids derived from natural gas. They have API gravities usually > 400. They concluded that in the future, the gas reserves of the delta will become a super World-Class Gas province.

Joseph et al., (2008) integrated spectral decomposition, AVO analysis, seismic attributes, principal component analysis, supervised neural facies classification and waveform calibration to delineate potential hydrocarbon bearing zones in plays from western Canada. Spectral Decomposition was used to identify erosional remnants of thin reservoirs in a meandering channel system by breaking down the seismic signal into its frequency components. In comparison a convectional amplitude sheet of the seismic data cube was produced but could not vividly reveal the thin channel reservoir as that spectral decomposition did.

Mohammed A. Eissa et. al., (2009) did a detailed study of thin, porous, gas saturated sandstone packages in the onshore Guajira Basin of Northern Colombia that includes petrophysical analysis, fluid substitute, and AVO analysis. The study shows that these thin sandstone reservoir are heterogenous in terms of rock properties, thickness and areal distribution and to reduce the risk of drilling dry wells, they recommended that better imaging of those thin sandstones can be achieved either by acquiring new seismic data using new techniques with better frequency content or by processing existing seismic data for frequency enhancement.

John P. Castagna et. al., (2002), He worked on how Spectral Decomposition can be used as a hydrocarbon Indicator. Through the various case studies he performed, he could identify three distinct spectral hydrocarbon indicators that are best revealed by proper spectral decomposition, these are; 1. Abnormal seismic attenuation 2. Low frequency shadows with hydrocarbon related bright spots and 3. Differences in "Tunning" frequency between gas and brine sands.

CHAPTER 2

2.1   GEOLOGY OF THE NIGER DELTA

INTRODUCTION

The Niger Delta is situated in the Gulf of Guinea and extends throughout the Niger Delta Province as defined by Klett and others (1997). From the Eocene to the present, the delta has prograded southwestward, forming depobelts that represent the most active portion of the delta at each stage of its development (Doust and Omatsola, 1990). These depobelts form one of the largest regressive deltas in the world with an area of some 300,000 km^3 (Kulke, 1995), a sediment volume of 500,000 km^3 (Hospers, 1965), and a sediment thickness of over 10 km in the basin depocenter (Kaplan and others, 1994).

The Niger Delta Province contains only one identified petroleum system (Kulke, 1995; Ekweozor and Daukoru, 1994; this study). This system is referred to here as the Tertiary Niger Delta (Akata – Agbada) Petroleum System. The maximum extent of the petroleum system coincides with the boundaries of the province. The minimum extent of the system is defined by the areal extent of fields and contains known resources (cumulative production plus proved reserves) of 34.5 billion barrels of oil (BBO) and 93.8 trillion cubic feet of gas (TCFG) (14.9 billion barrels of oil equivalent, BBOE) (Petroconsultants, 1996a). Currently, most of this petroleum is in fields that are onshore or on the continental shelf in waters less than 200 meters deep and occurs primarily in large, relatively simple structures. A few giant fields do occur in the delta, the largest contains just over 1.0 BBO (Petroconsultants, Inc., 1996a).

Among the provinces ranked in the U.S. Geological Survey's World Energy Assessment (Klett and others, 1997), the Niger Delta province is the twelfth richest in petroleum resources, with 2.2% of the world's discovered oil and 1.4% of the world's discovered gas (Petroconsultants, Inc. 1996a).

The onshore portion of the Niger Delta Province is delineated by the geology of southern Nigeria and southwestern Cameroon (Figure 2.1). The northern boundary is the Benin flank--an east-northeast trending hinge line south of the West Africa basement massif. The northeastern boundary is defined by outcrops of the Cretaceous on the Abakaliki High and further east-south-east by the Calabar flank--a hinge line bordering the adjacent Precambrian. The offshore boundary of the province is defined by the Cameroon volcanic line to the east, the eastern boundary of the Dahomey basin (the eastern-most West African transform-fault passive margin) to the west, and the two- kilometre sediment thickness contour or the 4000-meter bathymetric contour in areas where sediment thickness is greater than two kilometres to the south and southwest. The province covers 300,000 km^2 and includes the geologic extent of the Tertiary Niger Delta (Akata-Agbada) Petroleum System.

The deltaic sedimentation in the Niger Delta is considered to depend on rate of subsidence (Rs) and rate of deposition (Rd) and it is noted that the case of regression (Rd>Rs) is dominant with minor transgressive (Rs>Rd) phases. The interplay between these various conditions results in the development of various sedimentary megaunits called depobelts.

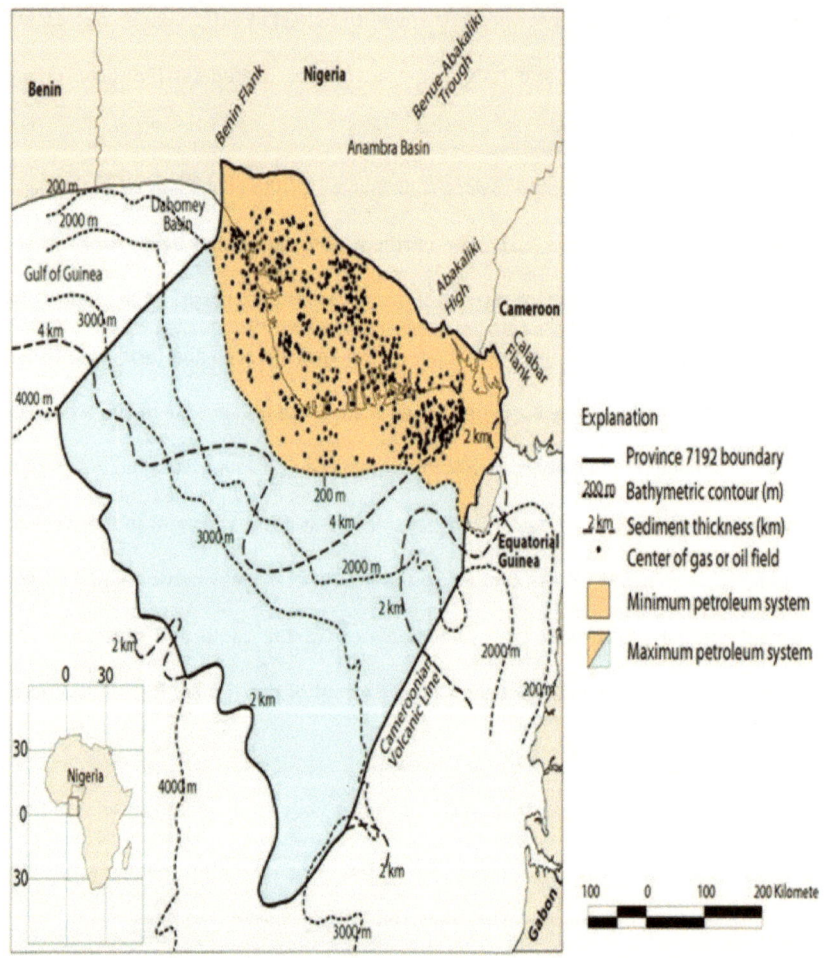

Figure 2.1: Index Map of Nigeria and Cameroon, Map of the Niger Delta Showing Province Outline (maximum petroleum system); bounding structural features; minimum petroleum system as defined by oil and gas field center points (data from petroconsults, 1996a); 200, 2000, 3000 and 4000 m bathymetric contours; and 2 and 4 km sediment thickness.

It is generally agreed that the modern Niger Delta is built on an oceanic crust. Supporting arguments come from the pre-continental drift reconciliation (Carey, 1991; Stoneley, 1966) – which indicates an important overlap of NE Brazil on the present Niger Delta; and from a series of geological and geographical observations e.g. the presence of a series of linear subdued and alternatively positive and negative anomalies beneath the Niger delta, interpreted by Burke *et al., (1971)* as seafloor spreading lineations (Mascle, 1976). The delta sequence is extensively affected by synsedimentary and post sedimentary normal faults which can be traced over considerable distances along strike.

The availability of many factors such as the right rock types serving as the source rock (shale) and reservoir rock (sandstone), temperature, means of transport (fault) which are favorable for the generation, migration, accumulation and retention of hydrocarbon in the Niger delta, making the delta very famous in terms of hydrocarbon occurrence.

2.2   DEVELOPMENT OF THE DELTA

Three major cycles of sedimentation have been established in the Niger Delta as well as other parts of the southern Nigeria Sedimentary Basin. These are:

- Lower Creataceous to Santonian Cycle (Oldest),
- Campanian to Paleocene Cycle, and
- Paleocene/Lower Eocene to Date cycle (youngest)

The third sedimentary cycle, commencing in the Paleocene/Early Eocene, is responsible for the main part of the delta's growth. The Niger delta oil province with its commercial oil fields is confined to the area covered by thick sequence of rocks belonging to the youngest (Tetiary) sedimentary cycle. (Short and Stuable, 1967).

2.3 THE PRESENT NIGER DELTA

The history of the delta since its inception in the Paleocene/Early Eocene is one of a major regression with gradual southward offlap of such macrolenses (Frankl and Cordry, 1967).

Thus, the sequence starting with coarse sandy deposits and ending with marine clays, is not only observed laterally, but is encountered also vertically in the Niger delta. In cross-section a time stratigraphic unit of such deltaic sediments is characteristically S-shaped or sigmoidal (Merki, 1972). The formations are therefore strongly diachronous, their age becoming progressively younger in a downdip direction and ranging from Paleocene to Recent (Figure 2.2).

2.4 SUBSURFACE FORMATIONS OF THE TETIARY NIGER DELTA

The Tertiary Niger Delta consists of an overall regressive plastic sequence about 9,000-12,000 m thick. The overall sequence is strongly dichronous (time transgressive). Three main rock-stratigraphic units had been proposed for the subsurface of the Niger Delta (Short and Stauble, 1967; Frankl and Cordry, 1967; Avbovbo, 1978).

The three sedimentary sequences established are:

- Akata Formation
- Agbada Formation
- Benin Formation

2.4.1 Akata Formation

This formation is characterized by a uniform shale development as evident in gamma and SP Logs. These prodeltaic shales are medium to dark grey, fairly hard, or at places soft, gumbo-like, and sandy or silty. The shales are undercompacted and may contain lenses of abnormally high pressured siltstone or fine grained

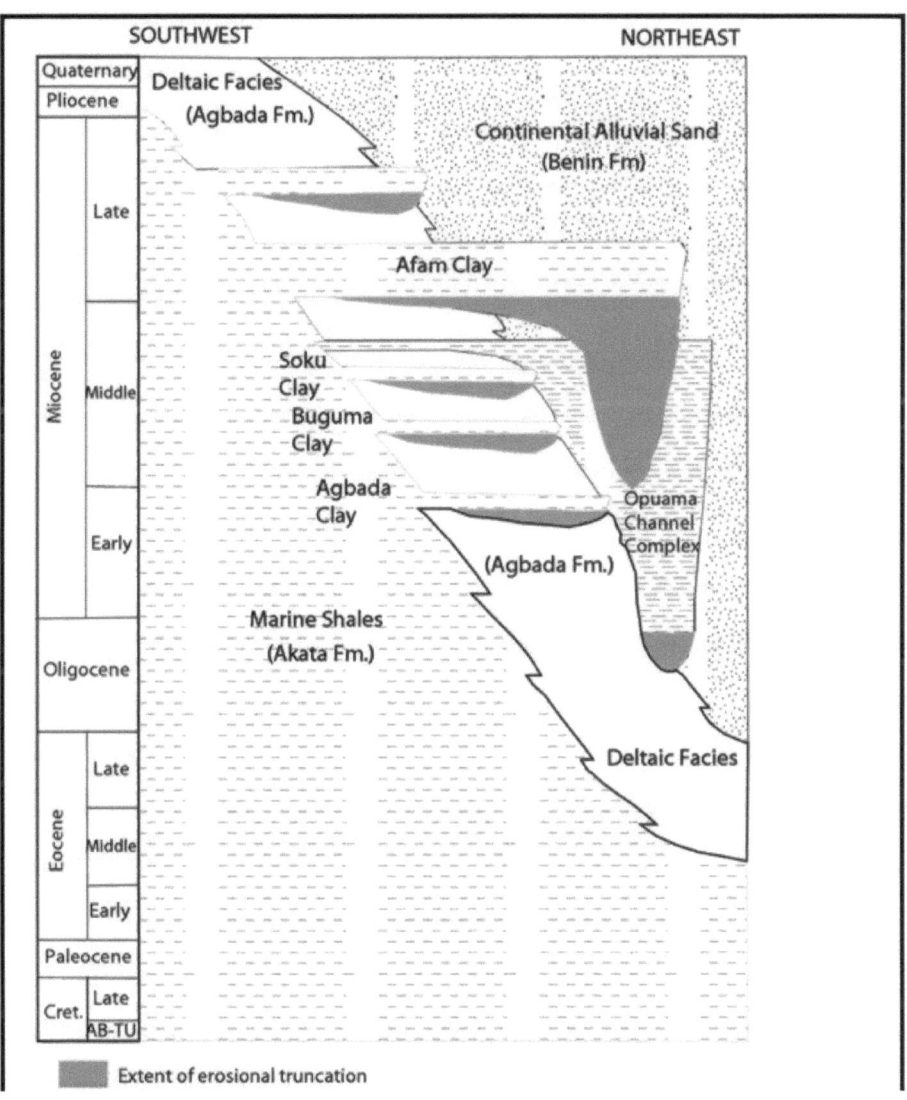

Figure 2.2: Stratigraphic columns showing the three formations of the Niger Delta. Modified from Shannon and Naylor (1989) and Doust and Omatsola (1990).

sandstone. Because most wells drilled in the Niger delta did not encounter the base of the Akata Formation, an estimation of its thickness is possible only in the northern part of the delta where the Formation has been drilled through into the cretaceous. The thickness of the sequence is not known for certain but may reach 7,000 m in the central part of the delta (Doust and Omatsola, 1990).

Generally, Akata Formation contains a rich foraminiferal fauna. Planktic foraminifera may constitute more than 50% of the microfana. The benthonic foraminifer's assemblage indicates deposition of the shale on a shallow marine shelf. The known age of the formation ranges from Eocene to Recent. However, it may be as old as Paleocene. Hydrocarbon migrates vertically along growth fault to accumulate in sandstone reservoirs of the Agbada Formation (Avbovbo, 1978)

2.4.2 Agbada Formation

This Formation consists of an alternating sequence of sandstones and shales of delta-front, distributary channel, and deltaic plain origin. The alternating sequence of sandstones and shales of the Agbada Formation has been shown to (Weber, 1971) be cyclic sequence of marine and fluvial deposits. The sandstones are medium to fined-grained, fairly clean and locally calcareous, glauconitic, and Shelly. They consist mainly kaolinite (average value 75%') with small amounts of mixed-layer illite and montmorillonite. At the central part of the delta the formation attains a maximum thickness 3,940 meters and thins northward and toward the northwestern and eastern flanks of the delta. North of the Benin City- Onitsha- Calabar axis, the Agbada Formation is poorly developed or absent. The shales of this formation contain a microfauna which is best developed or absent. The shales of this formation contain a microfauna which is best developed at the base of individual shale units. The depth environment of these fossil assemblages ranges from littoral-estuarine to marsh types of fauna developed at a water depth of approximately 100 metres. The age of the Agbada Formation varies from Eocene to Recent. The Agbada formation is of most concern in exploration for hydrocarbon in southern Nigeria due to the presence of growth faults and rollover anticlines that help in trapping hydrocarbon.

2.4.3 Benin Formation

This formation consists of predominantly massive, highly porous, freshwater-bearing sandstones, with local thin shale interbeds which are considered to be of braided-stream origin. Mineralogically, the sandstones consist dominantly of quartz and potash feldspar and minor amounts of plagioclase. The sandstones constitute 70 to 100% of the formation. Where present, the shale interbeds usually contain some plant remains and dispersed lignite. The formation attains a maximum thickness of 1,970 metres at the Warri-Degma area, which coincides with that (i.e depocentre) of the Agbada Formation. Most companies exploring for oil in the Niger delta arbitrarily define the base of the Benin Formation by the deepest freshwater-bearing sandstone that exhibits high resistivity.

Short and Stauble (1967) defined the base of the Benin Formation by the first marine foraminifera within shales, as the formation is non-marine in origin. Avbovbo (1978) agrees with Short and Stauble *(op.cit.)* and has also demonstrated that the base of the freshwater in the delta sediments extends into the Agbada Formation and thus not coincident with the base of the Benin Formation. Composition, structure and grain size of the sequence indicate deposition of the formation in a continental probably upper deltaic environment. The age of the formation varies from Oligocene to Recent. In the subsurface of the eastern part of the Niger delta, a clay section, the "Afam Clay Member" is locally recognized. The member has the form of a canyon fill which strikes in a South – Southeast direction, from slightly north of Afam-1 well to the west of the Imo River estuary. A maximum thickness greater than 8,000 metres has been established for the member. No counterpart to the Afam Clay Member is known in the recent (modern) Niger Delta.

2.5 BASIN ARCHITECTURE

The area of the Niger delta can be divided into a series of Depobelts, separated by major synsedimentary fault zones (Figure 2.3). These Depobelts can be thought of as transient basinal areas succeeding one another in space and time as the delta prograded southward. Why the delta evolved in a punctuated fashion is not fully understood. Subsidence and sedimentation within a depobelt may have been facilitated by large-scale withdrawal and forward (seaward) movement of undercompacted and geopressured marine shales under the weight of the advancing Paralic clastic wedge. At a certain stage, however, it seems that further subsidence (and hence sedimentation) could no longer be accommodated and the focus of deposition shifted basinward to form a new depobelt. This seaward advance coincided with rapid advance of the coastline and a shift of fluvio-marine deposition to the new depobelt. At the same time synsedimentary and most postsedimentary faulting ceased within the abandoned depobelt, which was covered by alluvial sands and "fossilized". A depobelt therefore, forms the structurally and depositionally most active portion of the delta at each stage of its development. The magnitude of throw and growth on faults bounding Depobelts is usually such that much of the Paralic sequence on the downthrown side is younger than that on the upthrown side. On delta flanks where the Paralic sequence is thinner, this is not always the case. Growth faults affecting the sequence within Depobelts form the boundaries of macrostructures (or individual delta units), each with its own sand-shale distribution pattern and structural style. Depobelts (or megastructure as they were originally called) comprise, in fact families of genetically and temporally related growth-fault trend, or macrostructures. The distal portion of the depobelts is broad, southward-dipping regional flanks whose strata typically demonstrate considerable expansion into landward-dipping counter-regional faults of unknown throw. Six (6) depobelt has been recognized within the Niger delta basin, which are distinguished primarily by their age and are best differentiated in the central area (Figure 2.3).

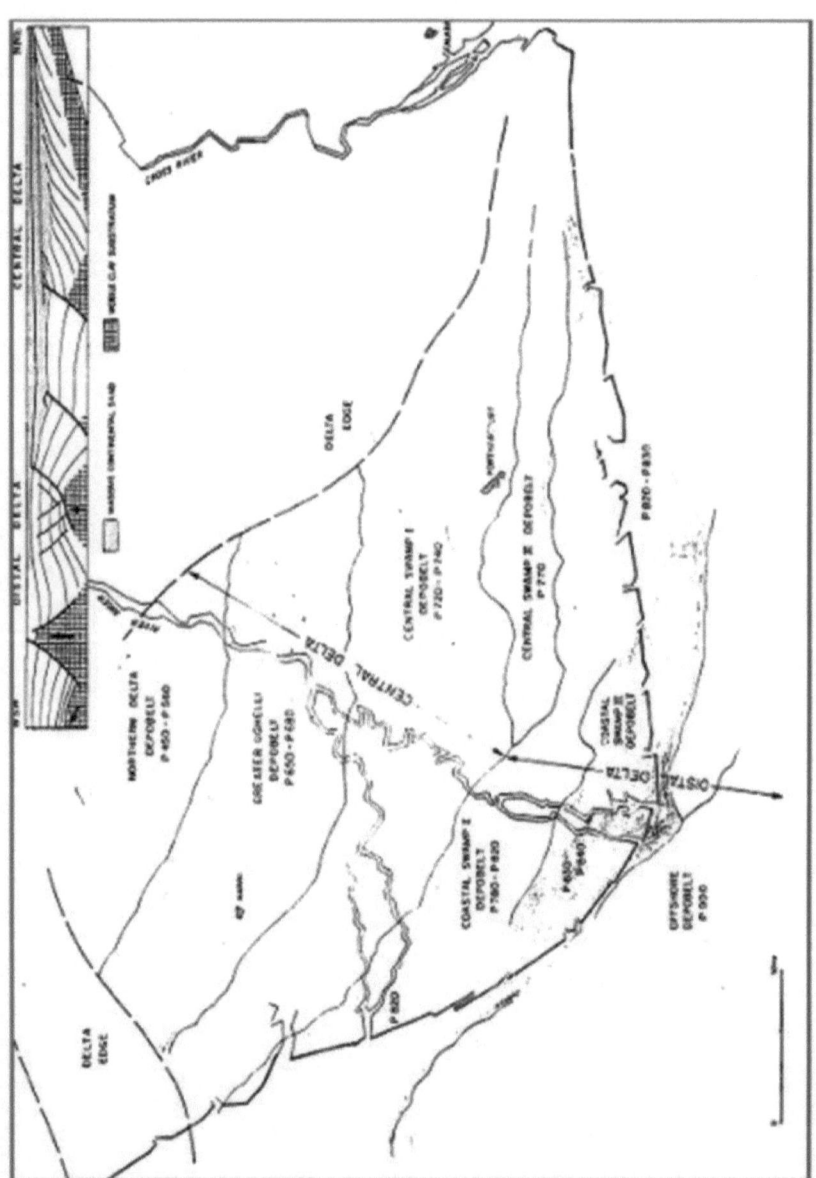

Figure 2.3: Stratigraphic Column Showing the Three Formation of the Niger Delta (Doust and Omatsola, 1990) Inset is the three distinguished areas - delta edge, central delta and the distal delta.

The depositional system in the Depobelts of the Niger delta basin has been described as "escalator regression" by Knox and Omatsola, 1989 characterized by rapid advance of alluvial sands and is due to cessation of subsidence in a depobelt and the continuation of sediment supply. The delta edge marks the beginning of the area of growth faulting, and the last (distal part) includes complex structures like collapsed crests, k-faulted flanks, and back-to-back areas.

2.6 SUBSURFACE STRUCTURES OF THE NIGER DELTA

The delta sequence is deformed by synsedimentary faulting and folding. Synsedimentary faults known as growth faults developed by gravity sliding during sedimentation. They deform the delta, affecting the Agbada and Akata formation. The growth fault give rise to different structural relationships and line up in actuate trends which are sub parallel to the coastline.

The subsurface structures of the Niger Delta originate because the great mass of marine clays of the Akata formation, which underlies the fluviomarine and fluviatile deposits of Agbada and Akata Formations is undercompacted and overpressure. The clays contain free water and their bulk density is lower than the density of the overlying sands and compacted shales of the Benin and Agbada formations. Differential loading of this "clay substratum" has created gravitational instability to which the mobile clays reacted by lateral and upward flowage. This mechanism occurs in the subsurface of the Niger Delta. Affected by the deformation are the Akata and Agbada formations. The deposits are only affected by regional tilting due to late subsidence of downdip parts of the delta. The structures observed are growth faults and rollover anticlines associated with these faults on their downthrown side.

2.6.1 GROWTH FAULT

The most common of the subsurface structural phenomena on seismic reflection profiles is the growth fault. A growth fault can be defined as a fault that offsets an active surface of deposition (Merki, 1972). It is characterized by thicker deposits in the downthrown block relative to the upthrown block. Deltaic growth faults differ from normal, 65^0 shear faults in the attitude of their fault planes. The main strike direction of these faults trends is roughly parallel to ancient delta shorelines and seems to be independent of the pattern of earlier fractures (i.e., the faults are not induced by basement tectonics). The growth fault planes exhibit a marked flattening with depth as a result of compaction.

Growth faults are the most common type of deformational feature in the Niger Delta. They are generally formed at the same time of deposition and are active throughout the sedimentation. Laterally, growth faults have a crescent shape, usually concave towards the downthrown block. The seaward flank of the clay ridge is the natural location for early growth fault (Short and Stauble, 1967).

2.6.2 ROLLOVER ANTICLINES

Rollover anticlines develop in the downthrown block of growth faults as a result of the warping of the hanging wall fault block towards the fault. Due to the curved nature of these faults, there is downward movement of sediments in the downthrown block thus creating a dip towards the fault (Figure 2.4).

It has been demonstrated that the Niger Delta faults have curved, concave-upward fault planes. Movement along these fault planes results in warping of the sediments in the downthrown block in the fashion of a reversed drag along an axis parallel to the fault. This creates a so-called

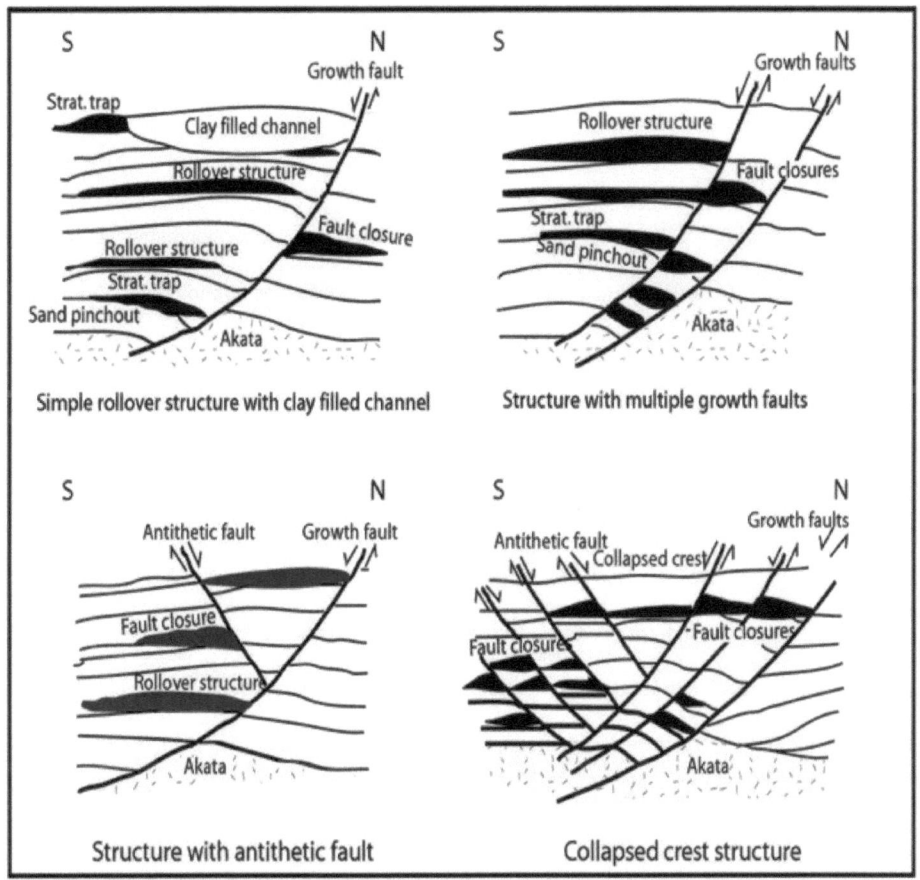

Figure 2.4: Examples of Niger Delta oil Field Structures and Associated Trap types. Modified after Doust and Omatsola (1990) and Stacher (1995).

"rollover", i.e. dip into the fault may only be introduced at a later stage as a result of regional tilting. Gentle undulations along the axis of the structure provide closure in directions parallel to the fault (Merki, 1972).

2.6.3 TRAPS AND SEALS

A Trap is a geometric configuration in the subsurface that is capable of containing oil and gas. According to Trapartite classification of Sanders (1943), traps are classified into four which are

- Structural traps
- Stratigraphic traps
- Combination traps
- Hydrodynamic traps

Stratigraphic traps result from lateral variation in lithology of the reservoir rock or a break in its continuity, while structural traps are those whose upper boundaries are concave when viewed from below due to some local deformation like faulting, folding or the combination of both. A combination of stratigraphic and structural trap is combination trap. Trapping mechanism in the Niger Delta can be structural, stratigraphic or a combination of both.

Most known traps in Niger Delta fields are structural although stratigraphic traps are not uncommon. The structural traps developed during synsedimentary deformation of the Agbada paralic sequence (Evamy and others, 1978; Stacher, 1995). As discussed earlier, structural complexity increases from the north (earlier formed depobelts) to the south (later formed depobelts) in response to increasing instability of the under-compacted, over- pressured shale. Doust and Omatsola (1990) describe a variety of structural trapping elements, including those associated with simple rollover structures, clay filled channels, structures with multiple growth. On the flanks of the delta, stratigraphic traps are likely as important as structural traps (Beka and Oti, 1995). In this region, pockets of sandstone occur between diapiric structures. Towards the delta toe (base of distal slope), this alternating sequence of sandstone and shale gradually grades to essentially sandstone.

Another type of stratigraphic trap encountered in the Niger Delta petroleum province is of the Barrier Bar type. Barrier bars are generated by long shore currents and wave action. They are usually several kilometres wide, many kilometres (upward of 30km) long and parallel to the coast line. Their sand thickness is often few metres; in some less than 15 metres and have the character of wide spread sheet sands. With good shale breaks that can reduce vertical permeability, many of them in the Niger Delta form good barrier bar traps.

The primary seal rock in the Niger Delta is the interbedded shale within the Agbada Formation. The shale provides three types of seals—clay smears along faults, interbedded sealing units against which reservoir sands are juxtaposed due to faulting, and vertical seals (Doust and Omatsola, 1990). On the flanks of the delta, major erosional events of early to middle Miocene age formed canyons that are now clay-filled. These clays form the top seals for some important offshore fields (Doust and Omatsola, 1990).

2.7 TECTONICS OF THE NIGER DELTA

The tectonic framework of the continental margin along the West Coast of equatorial Africa is controlled by Cretaceous fracture zones expressed as trenches and ridges in the deep Atlantic. The fracture zone ridges subdivide the margin into individual basins. In Nigeria, it forms the boundary faults of the Cretaceous Benue-Abakaliki trough, which cuts far into the West-African shield. The trough represents a failed arm of a rift triple junction associated with the opening of the South Atlantic. In this region, rifting started in the Late Jurassic and persisted into the Middle

Cretaceous (Lehner and De Ruiter, 1977). In the region of the Niger Delta, rifting diminished altogether in the Late Cretaceous

After rifting ceased, gravity tectonism became the primary deformational process. Shale mobility induced internal deformation and occurred in response to two processes (Kulke, 1995). First, shale diapirs formed from loading of poorly compacted, over-pressured, prodelta and delta-slope clays (Akata Formation) by the higher density delta-front sands (Agbada Formation). Second, slope instability occurred due to a lack of lateral, basinward, support for the under-compacted delta-slope clays (Akata Formation). For any given depobelt, gravity tectonics were completed before deposition of the Benin Formation and are expressed in complex structures, including shale diapirs, roll-over anticlines, collapsed growth fault crests, back-to-back features, and steeply dipping, closely spaced flank faults (Evamy and others, 1978; Xiao and Suppe, 1992). These faults mostly offset different parts of the Agbada Formation and flatten into detachment planes near the top of the Akata Formation.

CHAPTER 3

3.1 MATERIALS AND METHOD OF STUDY

The materials used for this study includes:

- **Base Map:** The base map supplies information on the location of the study area
- **Well Logs:** Two well logs comprising of gamma ray log and porosity logs were used
- **Checkshot Data:** The check shot data was used in the conversion of time values to depth, and for tying well log to seismic at the reservoir of interest
- **Seismic Sections:** A set of seismic sections were utilized for this work.

3.2 PHYSICAL BASIS OF REFLECTION SEISMOLOGY

Reflection seismology starts with some acoustic pulse (a "bang") that generates an expanding wavefront. The bang is located at some elevation "A" (ground surface, water surface, etc.). At any given point along the expanding wavefront, we can imagine a raypath that is perpendicular to the wavefront. The wavefront will expand until it reaches some interface, here located at depth "B", which causes some of the energy to be reflected back to the surface where it can be recorded. What is physically measured by the recording instruments (located back at the "surface") are : a) the strength of the reflected energy, and b) the time it takes for the energy to travel from the surface down to the reflecting horizon, then back up to the surface again. This time is referred to as the two-way traveltime or "TWT". In principle, if we measure the TWT at many points along an interface, we can get a picture of the relief on that interface - echosounders are a good example of this process.

3.2.1 P AND S WAVES

When a solid body is disturbed by something such as an explosion, the disturbance propagates through the body as waves. There are a variety of different types of waves. Some travel only at interfaces between two different media and are called surface waves. Waves on the ocean surface are of this type. Other types of waves propagate through the solid body itself. These are called body waves. There are two different types of body waves, namely P and S waves that are generated by earthquakes. The "P" stands for "primary" and the "S" stands for "secondary" since that is the order in which the waves are recorded by seismographs.

The reality is that these two types of waves correspond to two different types of disturbances. The faster P waves involve changes in volume (compression) and the slower S waves involve shear motions. As such, they can be referred to as Compressional and shear waves respectively. In reflection seismic work (e.g., 3-D data), it is nearly always compressional waves that are generated and recorded, although more expensive shear wave surveys are becoming more popular.

There are certain terms that can be employed to describe any wavelet. For a simple sinusoidal wave, although wavelets generated by seismic sources are far from being simple sinusoids. The first aspect of a wave we might wish to describe is the wavelength (λ) which is a measure (in feet or meters) of the distance between successive repetitions of the waveform. The frequency (f) is the number of waveforms that will pass by a given point per unit time. Frequency is measured in cycles per second, or Hertz. Finally, the amount of displacement from a resting position is called the amplitude of the wave. Amplitudes can have positive or negative values, and for seismic interpretation the absolute range of amplitudes we see in a seismic record depends on how the data were scaled. Positive amplitude values are referred to as peaks, negative amplitudes are referred to as troughs (Figure 3.2). The reflections of seismic data are recorded and (usually) displayed as traces that show variations in amplitude as a function of time. That is to say that the Z axis of a seismic profile is a measure of time (TWT). If we can count the number of peaks (or troughs) in a given time interval, we can estimate the dominant frequency of the data in that interval. Given a particular frequency for an interval, and having an estimate of velocity (from a sonic log, "intuition", or some other source) for that interval, the wavelength can be derived from:

$$\lambda = \frac{v}{f}$$

Where v is the velocity of sound in the rock/sediment, λ is the wavelength, and is the frequency.

3.2.3 REFLECTION AND PHYSICAL PROPERTIES OF ROCKS

Acoustic energy is reflected where there is a change in acoustic impedance (AI) of two adjacent rock layers (Figures 3.1 and 3.3). The acoustic impedance is the product of a rock's velocity (V) times its density (ρ):

$$AI = \rho V$$

At such an interface the energy within an incident seismic pulse is partitioned into transmitted, reflected, refracted, and diffracted pulses. The relative amplitude of each of which depends on the velocities and densities of the two layers, and the angle on incidence on the two layers (Figure 3.3)

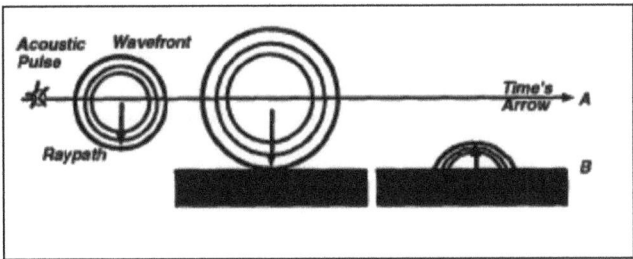

Figure 3.1: Simple Schematic Diagram Illustrating the Underlying Principles of the Seismic Method. One Wishes to Define Something About The Surface "B" At Some Depth Below the Interface "A" (The Land Surface or Water Surface).

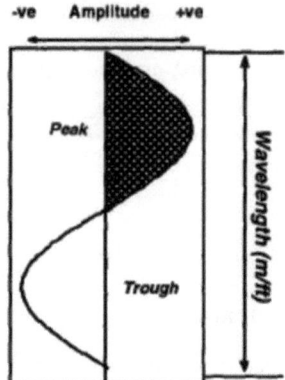

Figure 3.2: Definition of a Wavelet.

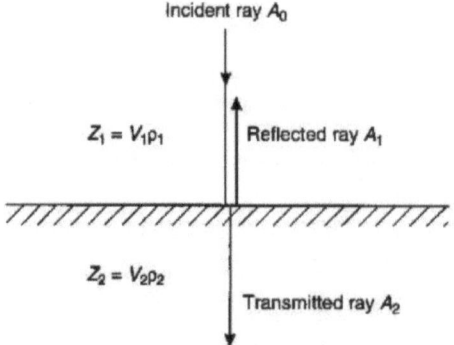

Figure 3.3: Reflected and Transmitted Rays Associated With a Ray Normally Incident on an Interface of Acoustic Impedance Contrast.

From figure 3.3 above, A_0 is the amplitude of the normally incident ray, A_1 is the amplitude of the reflected ray, and A_2 is the amplitude of the transmitted ray which travels through the interface to a second layer. The relative proportions of energy transmitted and reflected are determined by the acoustic impedance contrast, Z across the interface, and is given as the product of the rock density (ρ) and its wave velocity (V). Hence, the harder a rock, the higher its acoustic impedance contrast.

The reflection coefficient is a measure of the of acoustic impedance contrast across an interface on wave propagation, and it's given as the ratio of amplitude A_1 of the reflected to that of A_0 of the incident ray:

$$R = {A_1}/{A_0} = \frac{Z_2 - Z_1}{Z_2 + Z_1} = \frac{\rho_2 V_2 - \rho_1 V_1}{\rho_2 V_2 + \rho_1 V_1}$$

Where, ρ_1, V_1, Z_1 and ρ_2, V_2, Z_2 are the density, P-wave velocity and acoustic impedance values in the first and second layers, respectively.

Also, the transmission coefficient, T is the ratio of A_2 to A_0 and can be derived from the zoeppritz equation as:

$$T = {A_2}/{A_0} = \frac{2Z}{Z_2 + Z_1}$$

3.2.4 SEISMIC ACQUISITION

Seismic data acquisition involves the transmission of acoustic energy through the earth and receiving of the reflected signal by the appropriate detector on the earth surface. These sources can either be initiated on the surface or some few metres into the subsurface, though onshore seismic source are usually initiated in the subsurface. The major aim of seismic data acquisition is to provide subsurface seismic data of a prospect area and get a detailed geological description of the area.

Seismic data acquisition can be primarily classified into:

1) Two dimensional (2-D) seismic method,
2) Three dimensional (3-D) seismic methods.

3.2.4.1 SEISMIC ENERGY SOURCES

To generate reflections, we need some source of acoustic energy. The choice of what source to use will be a function of several variables. These include whether the seismic data are being collected at sea or on land, the depth and thickness of the principal targets of interest, environmental concerns and, last but not least, the amount of money available for the project.

There are a wide variety of seismic sources characterised by varying degree of energy levels and frequency spectrum. The various seismic sources in common use can be classified into 2 categories:

- Explosives
- Non-explosives

3.2.4.2 SEISMIC ENERGY DETECTORS

Two types of devices are used to record the reflected energy. At sea, hydrophones are towed behind a ship and convert pressure changes (from the reflected acoustic pulse) into electrical energy that can be recorded digitally. On land, geophones are implanted into the ground to convert ground motions (from the reflected acoustic pulse) into electrical energy. Similar to shot points where source arrays are deployed, several receivers are typically deployed at each "receiver location" (sometimes referred to as a "group") in an effort to boost the signal-to-noise ratio.

3.2.5 SEISMIC DATA PROCESSING

Figure 3.4 illustrates a simplified processing flow for seismic data. The processing flow for any given data set will be different from this simplified example, and may include steps not considered here.

3.2.6 SEISMIC DATA INTERPRETATION

3.2.6.1 Interpretation Workflow

1) Collect all pertinent data and reports,
2) Scan records for polarity and static shifts etc.,
3) Scan through section (line by line), overview of data quality, structure, and Stratigraphy,
4) Tie well, and seismic data,

5) Pick horizons and faults (loop tying)

6) Seismic stratigraphic analysis,

7) Structural analysis,

8) Contouring and mapping,

3.2.6.2 Data preparation

At this initial stage, the seismic data were tested for its uniqueness as to whether the profiles represent a carbonate succession or siliciclastics. Previous data and work done in the study area were collected and analyzed for efficient and effective interpretation. Knowing the stratigraphic and structural succession from the geology of the Niger delta before starting the interpretation helps in knowing what to look for and possibly where in the seismic data.

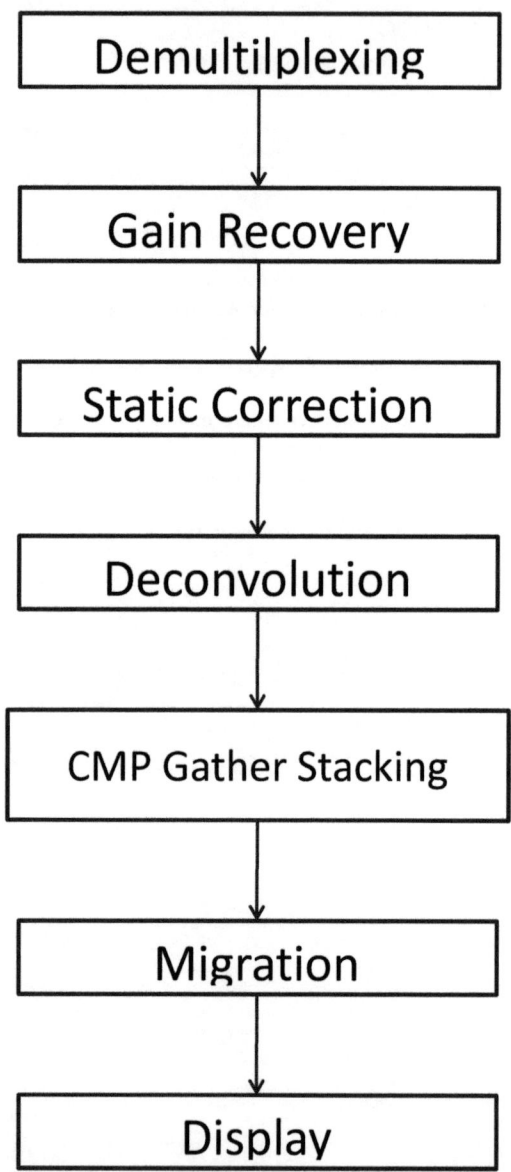

Figure 3.4: Simplified Processing Flow (After Yilmaz 1987)

Knowledge of the areas tectonic history helps to more quickly identify fault orientations and styles. Knowledge of what type of data is available (e.g., logs, core, biostratigraphy, production etc) help to determine what types of analyses to be undertaken.

The data was then given a quick "going over" in an attempt to determine the polarity, and the static shifts that have been applied. The seismic data has more than one vintage, each with a different static shift, display polarity. Access to digital data and workstation help minimize these differences interactively with time. This initial "reconnaissance" work provides an overview of the Stratigraphy and structural framework of the study area, as well as the data quality by examination across each seismic line.

At the end of this phase the key objectives of the study was defined and some working hypotheses to be tested during the interpretation were acquired.

3.2.6.3 FAULT PICKING

The term fault describes the displacement of a body of rocks by shearing or fracturing along a planar surface known as fault plane (Mcquillin et al; 1984). Faults are picked on the inlines (dip lines) on seismic sections. They can show up as offset of reflecting horizons with various breaks following a slanting path on the sections. The various faulting on seismic sections are associated with the following features:

- Displacement of events
- Changes if events
- Discontinuities in reflection patterns
- Divergence in dip of horizons
- Changes in reflection pattern and disappearance of the reflection below suspected fault lines.

3.2.6.4 HORIZON MAPPING

Horizons are individual reflections which are map-able isochronous sedimentary units (Tearpock and Bischke; 1991). It is identified by distinctive pattern which can be observed over a relatively large extent. Three horizons were picked based on prospectivity from the well logs available.

3.2.6.5 LOOP TYING

This is the process of transferring information or features on the lines to traces and vice-versa. In effect, the act of tying both the horizons and faults on seismic gridlines continually extends the surface and eliminates a number of possible surface configurations (Tearpock and Bischke; 1991). To form a loop around a seismic grid, the prospective horizon mapped on an inline was first transferred to a continuous crossline (strike line) with little or no tectonic activities. This crosslines is then used to transfer the reflection to the remaining inlines thereby closing the loop. Faults on the seismic lines were tied at the grid intersections. The importance of loop tying is as follows:

- To establish a relationship between the traces of the seismic lines.
- To project mapped horizons into areas where well control may not exist.

3.2.6.6 MAP GENERATION

Time-Depth map was generated for each of the horizons. The time map shows the variation of time across the field at a particular horizon. The depth maps were generated by converting the time map to depth map with the aid of the time depth curve. The fluid contacts from the well logs were identified on the depth maps and the respective prospect area for each zone was estimated.

3.3   WELL LOGGING

Developed by the schlumberger brothers (1978) in France, well logs are the workhorse of all reservoir characterization. Well log is the continuous measurement of a physical parameter down a borehole to provide data that are related to the petrophysical properties of the subsurface rocks which may be analyzed to monitor, simulate, and enhance a hydrocarbon reservoir. Petrophysics is defined as the study of the physical and chemical properties of rocks and the fluids they contain. Petrophysical log interpretation is one of the most useful and important tools available to a petroleum geologist. Besides their traditional use in exploration to correlate zones and to assist with structure and isopach mapping, logs help define physical rock characteristics such as lithology, porosity, pore geometry, and permeability. Logging data is used to identify productive zones, to determine depth and thickness of zones, to distinguish between oil, gas, or water in a reservoir, and to estimate hydrocarbon reserves. Also, geologic maps developed from log interpretation help with determining facies relationships and drilling locations. Petrophysicist uses rock properties and relationship to identify, quantify, and evaluate hydrocarbon reservoirs, source rocks, and seals. The principal deliverables are the static and dynamic reservoir description, subsurface fluid properties and distribution at and away from the wellbore.

The properties of primary concern are:

- Gas or oil – fluid type
- Percentage of pore spaces (occupied by fluids) per unit volume
- Percentage of pore space occupied by hydrocarbon – hydrocarbon saturation
- Reservoir thickness – net pay
- Rock transmissibility – permeability
- Rock type – Lithology and rock structure – dip, azimuth

3.3.1 BOREHOLE ENVIRONMENT AND INVASION PROFILES

Where a hole is drilled into a formation, the rock plus the fluids in it (rock-fluid system) are altered in the vicinity of the borehole. The rock and the fluid surrounding it are contaminated by the drilling mud, which affects logging measurements. Figure 3.5 is a schematic illustration of a porous and permeable formation which is penetrated by a borehole filled with drilling mud.

3.3.2 LOG TYPES AND USES

3.3.2.1 LITHOLOGY LOGS

Spontaneous Potential Log

The spontaneous potential (SP) log is used to identify impermeable zones such as shale, and permeable zones such as sand. The SP is a record of direct current (DC) voltage differences between the naturally occurring potential of a moveable electrode in the well bore, and the potential of a fixed electrode located at the surface. It is measured in millivolts (mV). Electric voltages arising primarily from electrochemical factors within the borehole and the adjacent rock create the SP log responses. These electrochemical factors are brought about by differences in salinities between mud filtrate and formation water within permeable beds. The salinity of a fluid is inversely proportional to its resistivity, and in practice salinity is indicated by

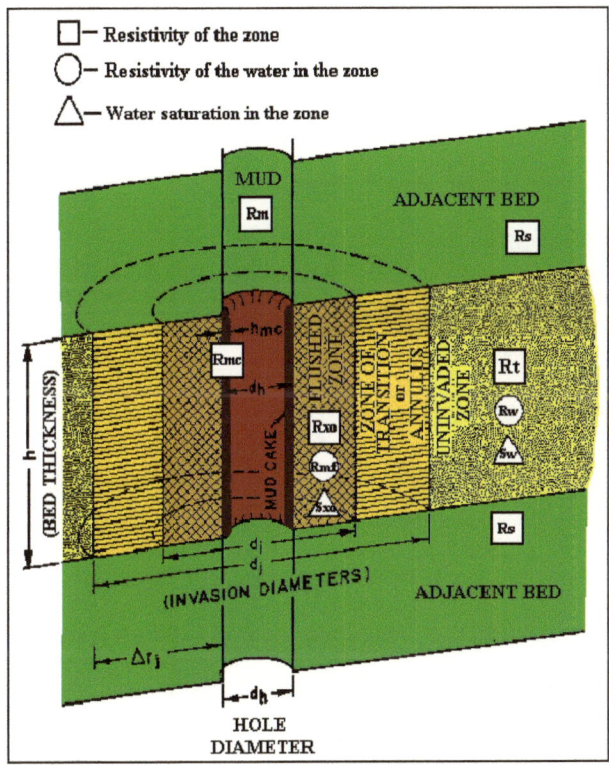

Figure 3.5: Borehole Environments and Invasion Profiles (www.petrolog.net/webhelp/logging_tools)

d_h –Hole diameter
R_m –Resistivity of the drilling mud
R_{mf} –Resistivity of mud filtrate
R_t –Resistivity of uninvaded zone (true resistivity)
R_{xo} –Resistivity of flushed zone
S_{xo} –Water saturation flushed zone
d_i –Diameter of invaded zone (inner boundary; flushed zone)
d_j –Diameter of invaded zone (outer boundary; invaded zone)
Δ_{rj} –Radius of invaded zone (outer boundary)

h_{mc} – thickness of mudcake
R_{mc} –Resistivity of the mudcake
R_s – Resistivity of shale
R_w –Resistivity of formation water
S_w –Water saturation of uninvaded zone

mud filtrate resistivity (R_{mf}) and formation water resistivity (R_w). Because a conductive fluid is needed for the generation of these voltages, the SP cannot be used in non-conductive (e.g., oil-base) drilling muds or in air-filled holes.

The SP log is recorded on the left hand of the chart and is used to:

1. Detect permeable beds,
2. Detect boundaries of permeable beds,
3. Determine formation water resistivity(R_w), and
4. Determine the volume of shale in permeable beds.

The SP response of shales is relatively constant and follows a straight line called a shale baseline (Schlumberger, 1989). The SP value of the shale baseline is estimated to be zero, and the SP curve deflections are measured from this baseline. Permeable zones are indicated where there is SP deflection from the shale baseline. Thus, if the SP curve moves to the left (negative deflection, $R_{mf} > R_w$) of the shale baseline, then permeable zones are present. The SP curve will not deflect from the shale baseline if $R_{mf} = R_w$ either in permeable or impermeable zones. The magnitude of the SP deflection depends on the difference between the mud filtrate and formation water resistivities, as a function of their differences in salinities, and not on the amount of permeability (Asquit and Gibson, 1982).

Gamma Ray Logs

Gamma ray logs measure natural radioactivity in formations and because of this measurement, they can be used for identifying lithologies and for correlating zones. Shale-free sandstones and carbonates have low concentrations of radioactive material, and give low gamma ray readings. As shale content increases, the gamma ray log response increases because of the concentration of radioactive material in shale however clean sandstone might also produce high gamma ray response if it contains potassium feldspar, micas, glauconite or uranium rich water. It is measured in API (American Petroleum Institute).

Gamma ray logs can be used to identify lithologies, correlating zones, and also for the determination of shale volumes. This volume is essential in calculating water saturations in shale –bearing Formations by shaly-sand analysis technique (Hilchie, 1978). The gamma ray log is particularly useful for defining shale beds when the SP readings are distorted (in very resistive Formations), when the SP is featureless (in freshwater-bearing Formations or in salty mud; i.e. $R_{mf} \approx R_w$) or when the SP cannot be recorded (in non-conductive mud, empty or air-drilled holes, and cased holes) (Schlumberger, 1989).

The GR log, unlike the SP log is not affected by the Formation water resistivity (R_w) because the GR log responds to the radioactive nature of the Formation rather that the electrical nature.

3.3.2.2 POROSITY LOGS

The three (3) main types of logs used to determine porosity of Formations are the Sonic, Density, and Neutron logs. The Density and the Neutron log are nuclear measurements, while the sonic log uses acoustic measurements. For all these devices, the tools' response is affected by the Formation porosity, fluids, and matrix. If the fluid and matrix effects are known or can be determined, the tools response can be related to porosity. All three logging techniques respond to the characteristics of the rock immediately adjacent to the borehole. Their depth of investigation is very shallow and therefore generally within the flushed zone (Schlumberger, 1989). They are used in concert to give accurate determination of porosity and a clear interpretation of Lithology.

Sonic Log

The sonic log is a porosity log that measures interval transit time (Δt) of a compressional sound wave traveling through one foot of formation. The sonic log device consists of one or more ultrasonic transmitters and two or more receivers. Interval transit time (Δt) in microseconds per foot ($\mu sec/ft$) or microseconds per metre ($\mu sec/m$) is the reciprocal of the velocity of a compressional sound wave in feet per second. The interval transit time (Δt) is dependent upon both Lithology and porosity. Therefore, a formation's matrix velocity must be known to derive sonic porosity. Modern sonic logs are Boreole Compensated (BHC) devices. These devices are designed to greatly reduce the spurious effects of borehole size variations, as well as errors due to the tilt of a tool with respect to the borehole axis (Schlumberger, 1972).

The very good vertical definition of the sonic log coupled with its reduced spurious effects because of borehole compensation makes the sonic log excellent for correlation (Schlumberger, 1989).

Neutron Log

Neutron logs are porosity logs that measure the hydrogen ion concentration in a formation. In clean formations (i.e. shale-free) where the porosity is filled with water or oil, the neutron log measures liquid-filled porosity (Φ_N, PHIN, or NPHI). Hydrocarbon in a porous Formation is concentrated in fluid-filled pores. Neutrons are emitted from a chemical source, collide with the hydrogen and experience an energy loss which is related to the Formation's porosity. The neutron log response is inversely proportional to porosity, so that low measurement unit values corresponds to high porosity and high measurement unit values corresponds to low porosity. A lowering of neutron porosity by gas is called "gas effect". Gas zones can be identified by comparing the neutron log with another porosity log or a core analysis. On the other hand, the presence of shales in a Formation leads to higher porosity than the actual Formation porosity. This happens because the hydrogen that is within the shale structure and in the water bound to the clay is sensed in addition to the hydrogen in the pore space. This increase in neutron porosity by the presence of shale is called "shale effect".

Formation Density Log

Density is measured in grams per cubic centimeter, g/cm^3 (or Kg/m^3 or Mg/m^3), and is indicated by the Greek letter ρ (rho). Two separate density values are used by the density log: the bulk density (ρb or RHOB) and the matrix density (ρma). The bulk density is the density of the entire formation (solid and fluid parts) as measured by the logging tool. The matrix density is the density of the solid framework of the rock.

This log is primarily used in porosity determination. The formation density log is a porosity log that measures electron density of a formation. It can assist the geologist to:

1. Identify evaporite minerals,
2. Detect gas-bearing zones,
3. Determine hydrocarbon density, and
4. Evaluate shaly sand reservoirs and complex lithologies (Schlumberber, 1972).

The density logging tool has a relatively shallow depth of investigation, and as a result, is held against the side of the borehole during logging to maximize its response to the formation.

The bulk-density curve (RHOB) is recorded in tracks 2 and 3 of well logs.

Resistivity Logs

Resistivity logs are electric logs which are used to:

1. Determine hydrocarbon versus water-bearing zones,
2. Indicate permeable zones, and

3. Determine porosity.

By far the most important use of resistivity logs is the determination of hydrocarbon versus water-bearing zones. Because the rock's matrix or grains are non-conductive, the ability of the rock to transmit a current is almost entirely a function of water in the pores. Hydrocarbons, like the rock's matrix, are also non-conductive; therefore, as the hydrocarbon saturation of the pores increases (as the water saturation decreases), the formation resistivity increases. A geologist by knowing or determining several parameters (a, m, n, and R_w) and by determining from Logs the porosity (Ø), formation bulk, or true resistivity (R_t), can determine the formation's water saturation (S_w) from the Archie's equation:

$$R_o = \frac{R_w}{\emptyset^m} \tag{3.37}$$

$$R_t = \frac{aR_w}{\emptyset^m S_w^n} \tag{3.38}$$

$$S_w = \left(\frac{F \times R_w}{R_t}\right)^{1/n} \tag{3.39}$$

Where: S_w is the Water saturation, F is the Formation factor (a/ $\emptyset^m$), a is the Tortuosity factor, m is the Cementation exponent, R_w is the Resistivity of formation water, R_t is the True formation resistivity as measured by deep reading resistivity log, and n is the Saturation exponent (most commonly 2.0).

Resistivity Logs can be classified into:
1. Induction tools: Uses coils to induce a current and measures the Formation's conductivity
2. Electrode tools (also called galvanic devices or laterologs): Have electrodes on the surface of the tool that emits currents and measures to resistivity of the Formation.

3.3.3 WELL LOG CORRELATION

The process of correlation of lithologic units in a stratigraphic sequence involves the use of various parameters, such as fossil content, lithologic facies, etc in mapping the lateral continuity and equivalence of these units. However this can also be achieved by using gamma ray log singly or in combination with some other logs that is descriptive of the characteristics of individual beds within a given stratigraphic sequence. The gamma ray log and the resistivity log were used to correlate lithologic units across four wells in this study.

The marker beds are first identified and correlated. Marker beds are lithologic units that are laterally extensive and visible in most part of the well. (They are shale units in this case). Once the marker beds have been established, the members of that stratigraphic sequence can then be correlated with respect to the trend of the marker beds. Well correlation is of particular importance because it allows for the deduction of:

- The presence of fault.
- The elevation of formations within a well relative to other wells.
- The geologic structures intersected by the wells.

3.4 SEISMIC ATTRIBUTES.

3.4.1 Introduction

The Oxford Dictionary defines an attribute as, *"A quality ascribed to any person or thing"*. This definition has been extended to: *"Seismic Attributes are all the information obtained from seismic data, either by direct measurements or by logical or experience based reasoning"*.

Since their introduction in the early 1970's, Complex Seismic Trace Attributes have gained considerable popularity, first as a convenient display form, and later, as they were incorporated with other seismically-derived measurements, they became a valid analytical tool for lithology prediction and reservoir characterization. The study and interpretation of seismic attributes provide us with some qualitative information of the geometry and the physical parameters of the subsurface.

It has been noted that the amplitude content of seismic data is the principal factor for the determination of physical parameters, such as the acoustic impedance, reflection coefficients, velocities, absorption etc. The phase component is the principal factor in determining the shapes of the reflectors, their geometrical configurations etc. One point that must be brought out is that we define all seismically-driven parameters as Seismic Attributes. They can be velocity, amplitude, frequency, and the rate of change of any of these with respect to time or space and we have developed a classification scheme for all attributes that is based on their computational characteristics. The principal objectives of the attributes are to provide accurate and detailed information to the interpreter on structural, stratigraphic and lithological parameters of the seismic prospect.

3.4.2 The Classification of Attributes

Attributes can be computed from prestack or from poststack data, before or after time migration. The procedure is the same in all of these cases. Attributes can be classified in many different ways. Several authors have given their own classification. Here we give a classification based on the domain characteristics of the attributes:

Pre-Stack Attributes

Input data are CDP or image gather traces. They will have directional (azimuth) and offset related information. These computations generate huge amounts of data; hence they are not practical for initial studies. However, they contain considerable amounts of information that can be directly related to fluid content and fracture orientation. AVO, velocities and azimuthal variation of all attributes are included in this class.

Post-Stack Attributes

Stacking is an averaging process which eliminates offset and azimuth related information. Input data could be CDP stacked or migrated. One should note that time migrated data will maintain their time relationships, hence temporal variables, such as frequency, will also retain their physical dimensions. For depth migrated sections, frequency is replaced by wave number, which is a function of propagation velocity and frequency.

Post-stack attributes are a more manageable approach for observing large amounts of data in initial reconnaissance investigations. For detailed studies, pre-stack attributes may be incorporated.

Attributes may be further classified by their computational characteristics:

Instantaneous Attributes

Instantaneous attributes are computed sample by sample, and represent instantaneous variations of various parameters. Instantaneous values of attributes such as trace envelope, its derivatives, frequency and phase may be determined from complex traces.

Wavelet Attributes

This class comprises those instantaneous attributes that are computed at the peak of the trace envelope and have a direct relationship to the Fourier transform of the wavelet in the vicinity of the envelope peak. For example, instantaneous frequency at the peak of the envelope is equal to the mean frequency of the wavelet amplitude spectrum. Instantaneous phase corresponds to the intercept phase of the wavelet. This attribute is also called the "response attribute". (Bodine, 1984).

These attributes may be sub-classified on the basis of the relationship to the geology:

Physical Attributes

Physical attributes relate to physical qualities and quantities. The magnitude of the trace envelope is proportional to the acoustic impedance contrast; frequencies relate to bed thickness, wave scattering and absorption. Instantaneous and average velocities directly relate to rock properties. Consequently, these attributes are mostly used for lithological classification and reservoir characterization.

Geometrical Attributes

Geometrical attributes describe the spatial and temporal relationship of all other attributes. Lateral continuity measured by semblance is a good indicator of bedding similarity as well as discontinuity. Bedding dips and curvatures give depositional information. Geometrical attributes are also of use for stratigraphic interpretation since they define event characteristics and their spatial relationships, and may be used to quantify features that directly assist in the recognition of depositional patterns, and related lithology.

Most of the attributes, instantaneous or wavelet, are a function of the characteristics of the reflected seismic wavelet. That is, we consider the interfaces between two beds. However, velocity and absorption are measured as quantities occurring between two interfaces, or within a bed.

Therefore, we can further sub-divide the attributes into two categories, as follows:

Reflective Attributes

Attributes corresponding to the characteristics of interfaces. All instantaneous and wavelet attributes can be included under this category. Pre-stack attributes such as AVO are also reflective attributes, since AVO studies the angle dependent reflection response of an interface.

Transmissive Attributes

Transmissive attributes relate to the characteristics of a bed between two interfaces. Interval, RMS and average velocities, Q, absorption and dispersion come under this category. Because individual attributes may be representative of several possible conditions we attempt to minimize this inherent uncertainty, or non-uniqueness, by combining multiple attributes in a logical fashion. Individual attributes measuring only one quantity are termed "Primitive" attributes. These primitive attributes may be logically, statistically or mathematically combined to form "Hybrid" attributes. The most common tool for performing this combination is through the use of Artificial Neural Networks.

3.5 Spectral Decomposition: A Powerful Risk Reduction Tool

Spectral decomposition is a novel seismic technique that was originally pioneered through research at BP and Amoco in the 1990's. Spectral decomposition is an imaging innovation that provides interpreters with high-resolution reservoir detail for imaging and mapping temporal bed thickness and geological discontinuities within 3D seismic surveys by breaking down the seismic signal into its component frequencies (Staffan Kristian Van Dyke – *"Spectral Decomposition: A Powerful Tool for the Seismic Interpreter" – Unpublished*).

A fully processed seismic survey contains all of the frequencies that are capable of being recorded by the geophones/hydrophones used for that particular survey (this is known as its "dynamic range"). After the seismic source has been "shot," the energy propagates downward into the subsurface and at each geologic boundary (e.g., an unconformity, bed boundaries, etc.), the seismic energy is reflected, refracted, and/or absorbed. As the wave front continues to propagate into the underlying sediments, it attenuates, causing the frequency content to decrease with depth, i.e., higher frequencies are better preserved at the top of the section. Due to this attenuation, the higher frequencies deeper in the seismic survey are "drowned" by the more dominant, lower frequencies. The purpose of spectral decomposition is to see the seismic response at different, discrete frequency intervals, as higher frequencies image thinner beds, while lower frequencies image thicker beds.

The concept behind spectral decomposition is that the seismic reflection from a thin bed has a characteristic expression in the frequency domain that is indicative of its thickness in time (Figure 3.6). For example, a simple homogeneous thin bed contains a predictable and periodic sequence of notches into the amplitude spectrum of the composite reflection (Praptono et al., 2003). However, typically a seismic wavelet contains the information from multiple subsurface layers and not just one simple thin bed. The combined seismic response from these multiple subsurface layers usually results in a complex tuned reflection which has a unique frequency domain expression; in order to help resolve these thin beds, spec-decomp can be used (Staffan Kristian Van Dyke – *"Spectral Decomposition: A Powerful Tool for the Seismic Interpreter" – Unpublished)..*

Spectral decomposition is used to break down the seismic survey into its component frequencies. When determining which frequencies to extract from the dataset, it's best to use a non-standard or octave scale in order to avoid potential harmonics (seeing the same information at multiples of its base frequency). Thus, multiple datasets are created at these pre-selected, discrete frequency intervals, e.g., 15.3 Hz, 29.6 Hz, 44.4 Hz, and so on (Figure 3.7). After determining these frequency intervals, each subsequent dataset produced via spec-decomp manifests *only* that particular frequency. After all datasets have been produced, the reservoir interval of interest can then be scrutinized in greater detail (Figure 3.8). This is carried out by capturing the seismic response at each frequency subset (15.3 Hz, 29.6 Hz, 44.4 Hz, etc.) – essentially, a "screen-capture" of the seismic image for each of these intervals can be input into an animated sequence from lower frequencies to higher frequencies, thus revealing spatial changes in stratigraphic thickness otherwise impossible to ascertain from the full frequency dataset. Spectral decomposition reveals details that no single frequency attribute can match.

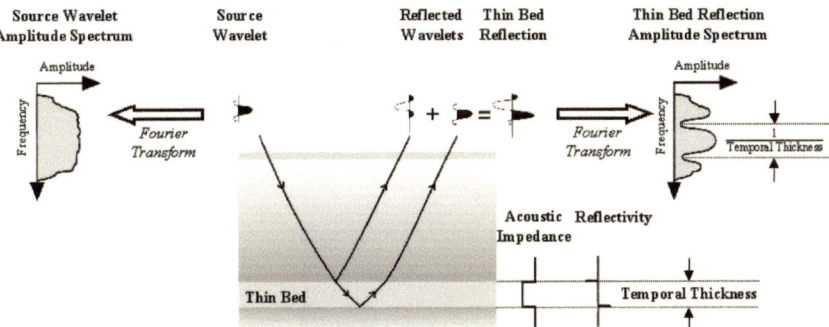

Figure 3.6: Spectral decomposition is used to identify thin beds through analysis of the frequency spectrum in a short window around the time of the bed. (Partyka *et al.*, 1999)

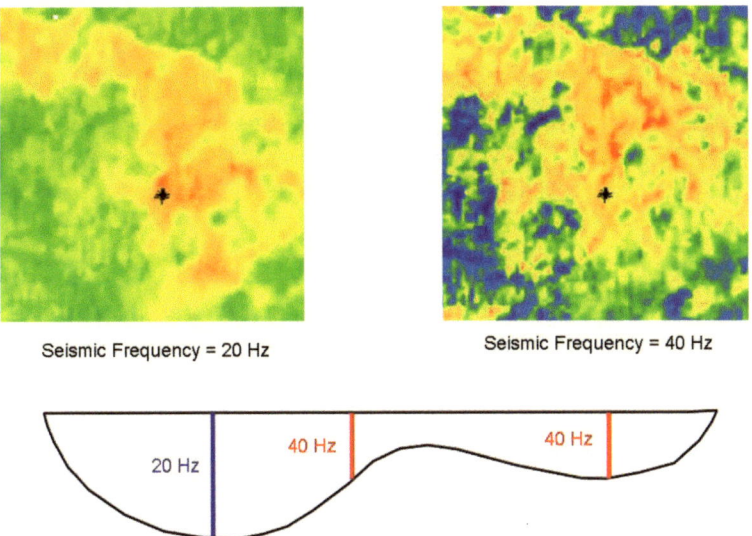

Figure 3.7: Comparison of Frequency Effects on thin Beds.

Note the different seismic response at 40 Hz as compared to 20 Hz; much more detail can be ascertained with the 40 Hz wavelet, however, the 20 Hz wavelet still manifests information about temporal bed thickness and the stratigraphic nature of the deposit.

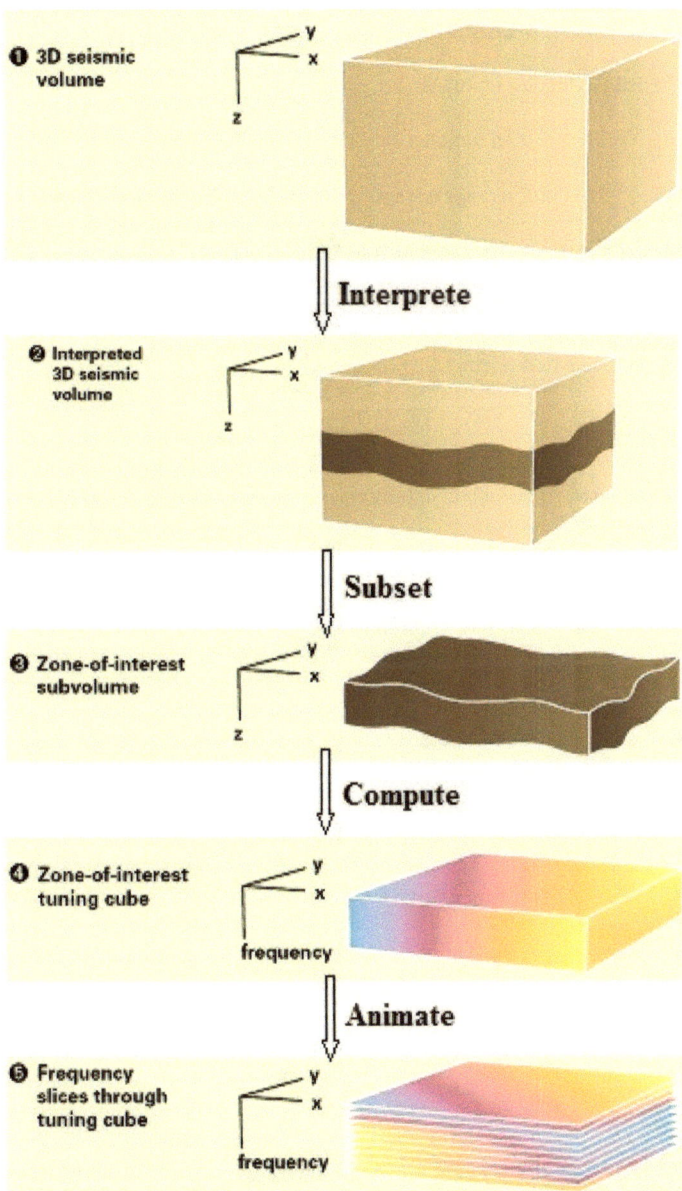

Figure 3.8: Spectral Decomposition Work Flow.

CHAPTER 4

4.1 DISCUSSION AND RESULTS PRESENTATION.

All Interpretation and analysis has been carried out on a 3-D Seismic data with inline range of 5800 to 6200, crossline range of 1480 to 1700. The Z – range is from 0 to 6000. Six wells were given in the well log data (Tmb 01, Tmb 02, Tmb 03, Tmb 04, Tmb 05 and Tmb 06) but four were used exclusively for this project. The used wells are Tmb 02, Tmb 05, Tmb 01 and Tmb 04. This was used because of their disposition on the base map and for easy correlation.

4.2 Conventional Mapping

As it has been stated in Chapter 1, two conventional Interpretations were employed in the course of the study and these are:

        a. Faults Interpretation and

        b. Horizon Mapping

Other analyses carried out are:

a. Well to Seismic Tie

b. Well Log Interpretation

c. Spectral Decomposition

- Thin Bed 1 (TB 01T)
- Thin Bed 2. (TB 03T)

d. Production of Maps.

4.2.1 FAULT MAPPING

Nine Faults were mapped in the seismic data. Two of which are major and it cuts through the whole volume of the cube. Some are neither major nor minor while some are completely classified as minor. The faults are; F1, F2, F3, F4, F5, F6, F7, F8, and F9. F1 and F9 are the major faults recognized in the seismic data while F8 and F5 are neither major nor minor. F2, F3, F4, F6, and F8 are minor faults, (Figure 4.1).

The faults are concentrated at the western part of the field. Although, Faults F8, and F7 are represented at the eastern part of the field aside the two afore mentioned major faults that cuts through the whole volume (F1 and F9). These two major faults cuts through the picked pay sand reflectors (TB 01T and TB 03T) and has been distinctively mapped by the frequency analysis outputs. Faults F7, F8, F1, F4 and F2 are growth faults while faults F5, F9, and F6 are antithetic faults dipping in an opposite sense to the growth faults (Figure 4.2).

Families of genetically and temporally related growth faults characteristic of the Niger Delta area are visible on the seismic section. The faults are generally normal faults, with listric geometries concaving seaward, and curvature ranging from linear to crescent-shaped which flattened with depth. These listric growth faults are syndepositional, exhibiting a significant increase in Stratigraphy thickening on the downthrown block or hanging wall evident by the bright and distinct signature on this side, and an increase in displacement with depth; which indicates that there were movements along the fault surface while the sediment was being deposited i.e. contemporaneous. The growth faults are formed by rapid sediment loading of the Agbada formation on the undercompacted, overpressured, and highly mobile Akata shale, while the antithetic faults are secondary faults which displays no growth and have a counter-regional dip, resulting from a conjugate compensation of overburden extension, explaining why they are generally shallow when compared to the growth faults.

In terms of relative ages of the faults visible on the seismic section, it can be established that the synsedimentary growth faults (F, G, and J) are the oldest, while the antithetic faults E, I, and K are the youngest.

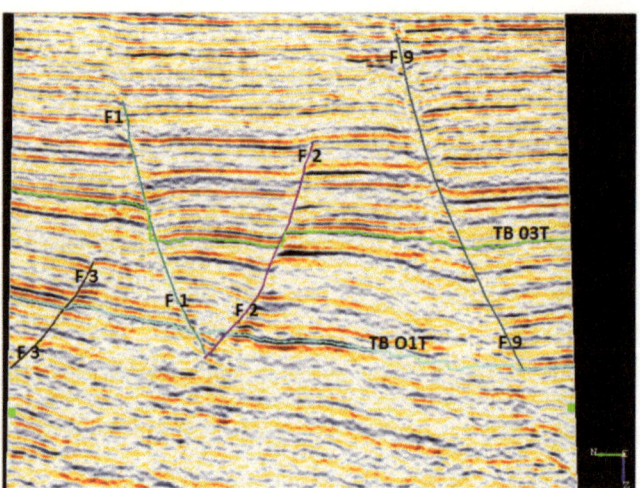

Figure 4.1: Some of the picked faults on inline 5915 with Mapped pay sand reflectors (TB 01T and TB 3T).

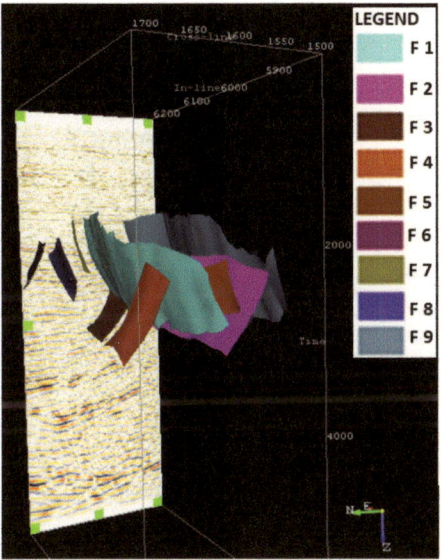

Figure 4.2: A view from the eastern part of the Volume; showing the mapped faults.

4.2.2 Horizon Tracking.

In this project two horizons were mapped, these are; TB 03T and TB 01T (Figure 4.3). In this case, two reflectors were tracked on the seismic sections. Well 5 was taken as the standard for the other wells, the makers TB 03T and TB 01T was interpreted on RokDoc and imported on the OpendTect for further processing.

An auto tracking was done for the picked reflectors. This then showed the lateral representation of the said reflectors (Figures 4.4 and 4.5). The auto tracking recognized the faults that cut through the reflector as it shows their respective discontinuities (Figure 4.4 and 4.5).

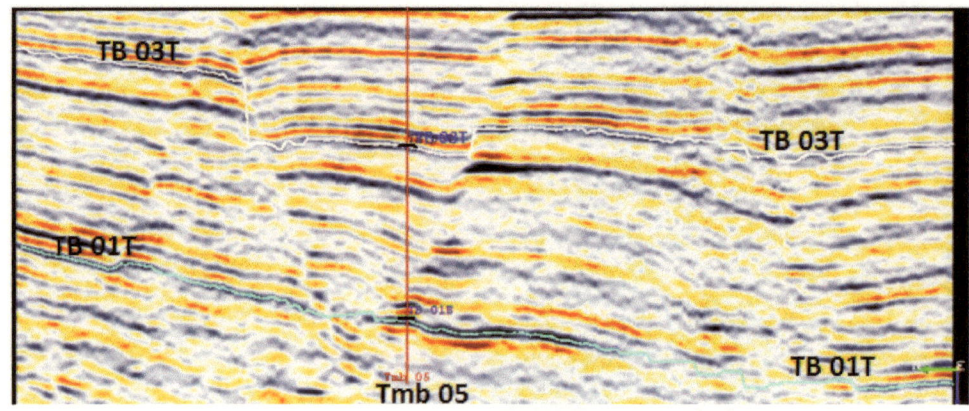

Figure 4.3: Two tracked reflectors across the two markers; TB 03T and TB 01B

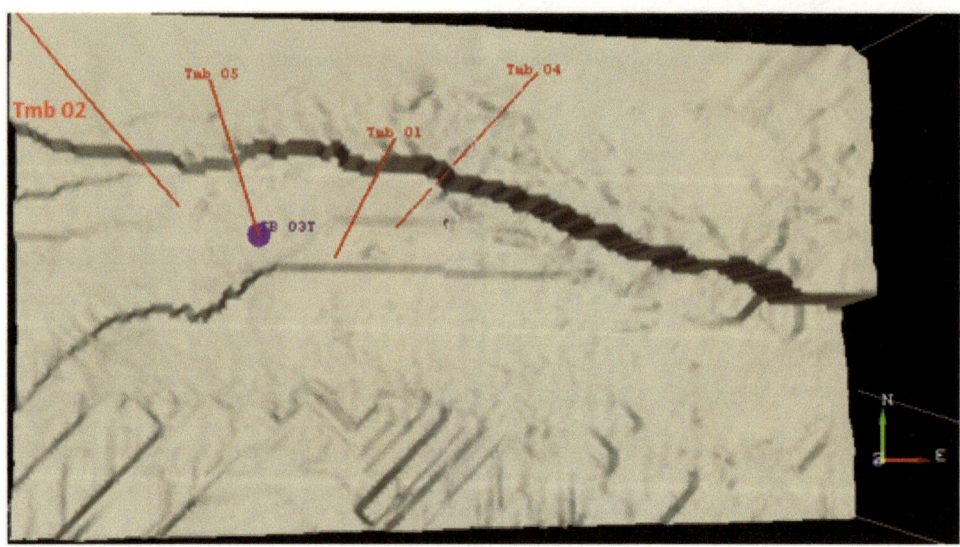

Figure 4.4: Reflector Horizon for TB 03T tracked across the volume with the used wells.

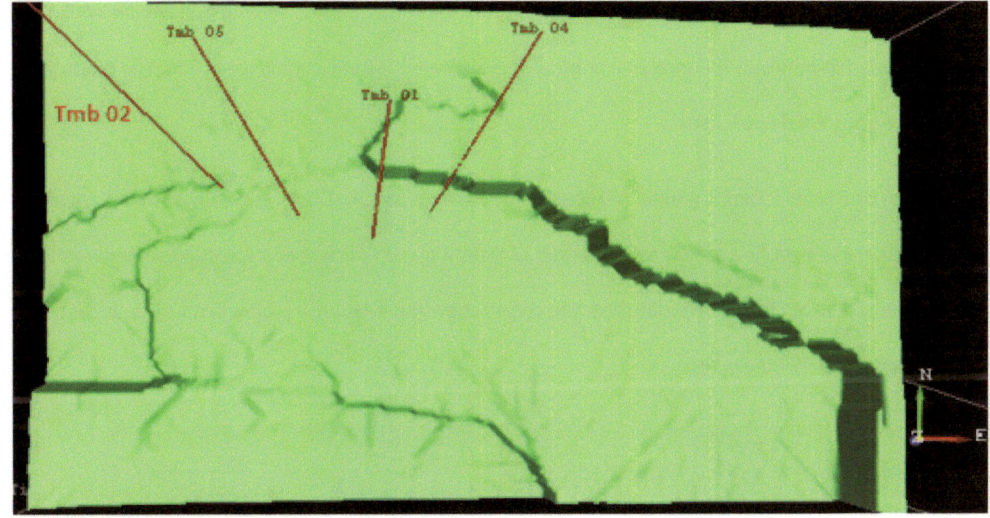

Figure 4.5: Reflector TB 01T tracked across the volume with the used wells.

4.3 Well to Seismic tying

The well to seismic tie was done using the OpendTect 4.2.0 Software. None of the wells given in the data used has both Density and Sonic Logs to a particular well. Gardner's Equation was used to convert the Density logs (i. e. RHOB) into Sonic Log (i. e. DT).

$$DT\_G = ((25/RHOB)^4)/140$$

Where: DT= Sonic Log

RHOB= Density Log

After computing the two logs, for wells 3, 4, 5, and 6; an importation was done into the software (i. e. OpendTect), and the proper tying was done. This was done by matching the synthetic wave form to the seismic wave form (Figures 4.6 and 4.7).

Data generated through the tying was used to plot the depth- time graph which shows the relationship between the time and depth of the surveyed seismic cube (Figure 4.8). The plotting was done using Excel application.

From the Time – Depth graphs (Figures 4.9, 4.10, 4.11 and 4. 12), it generally shows a linear and relatively straight line except for the graph of wells 4 and 5 which shows a conspicuous inward bend (Figures 4.10 and 4.11). In a nut shell, the curves show the relationship between time and depth as it concerns each of the wells.

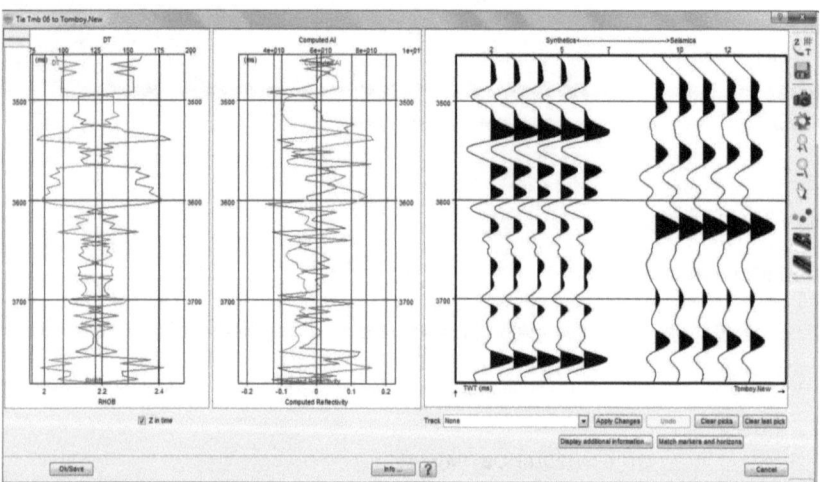

Figure 4.6: The Relationship between Sonic Log (DT), Computed Acoustic Impedance in Relation to Depth and the Generated Synthetic Wave Form and Seismic Wave Form.

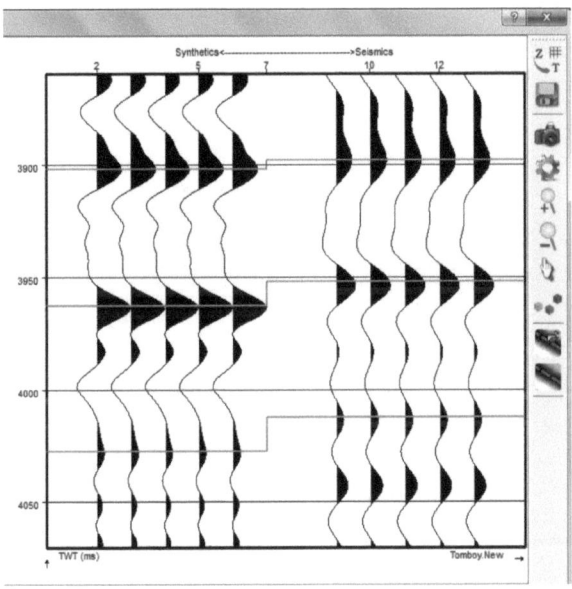

Figure 4.7: Matching of the Synthetic Wave Form to the Seismic Wave Form

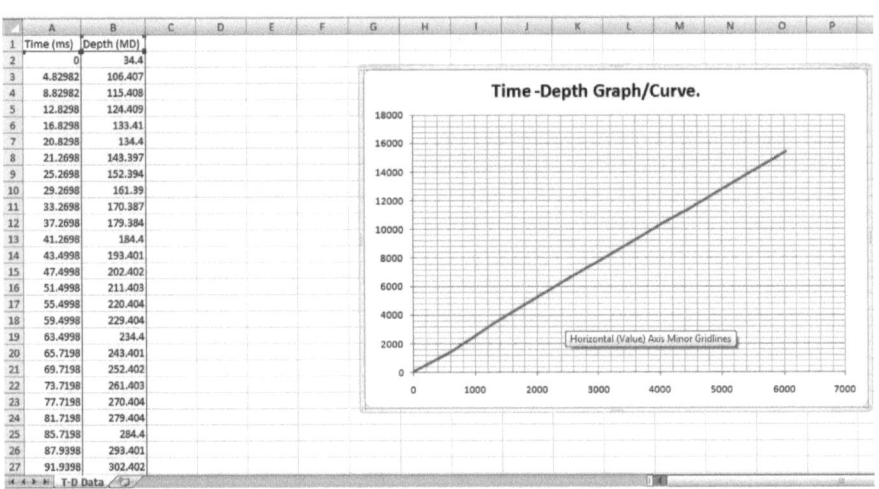

Figure 4.8: An Excel Spread sheet, Showing the Generated Data for one of the Four Wells used for tying and the Consequent Depth - Time Plot.

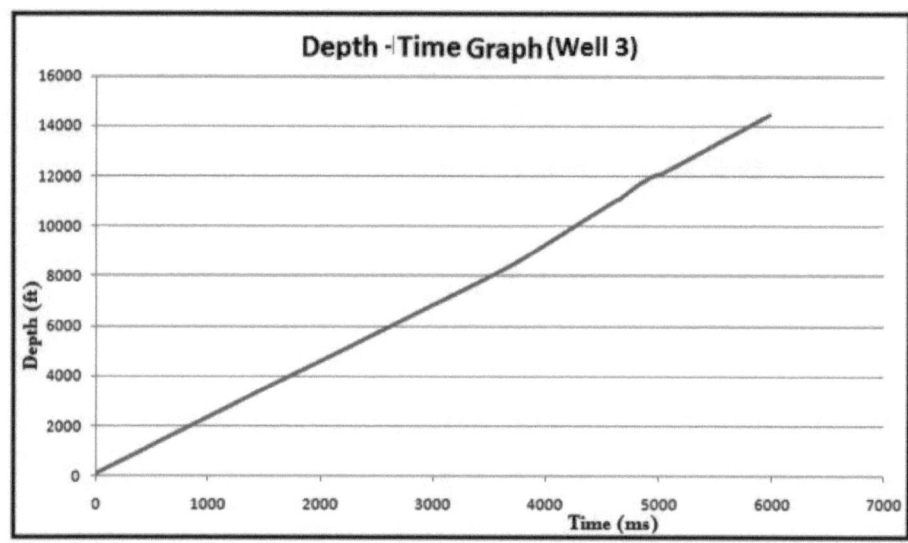

Figure 4.9: The Depth – Time Graph of Well 3.

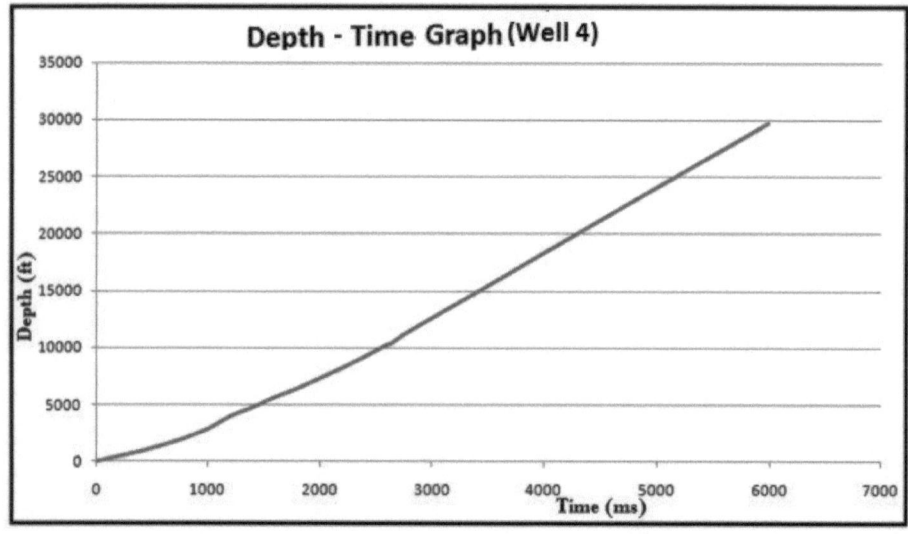

Figure 4.10: The Depth – Time Graph of Well 4.

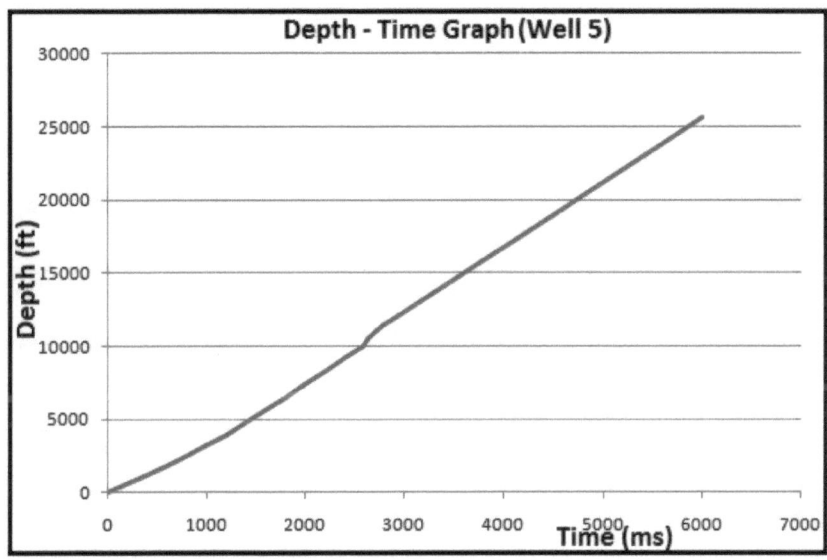

Figure 4.11: The Depth – Time Graph of Well 5.

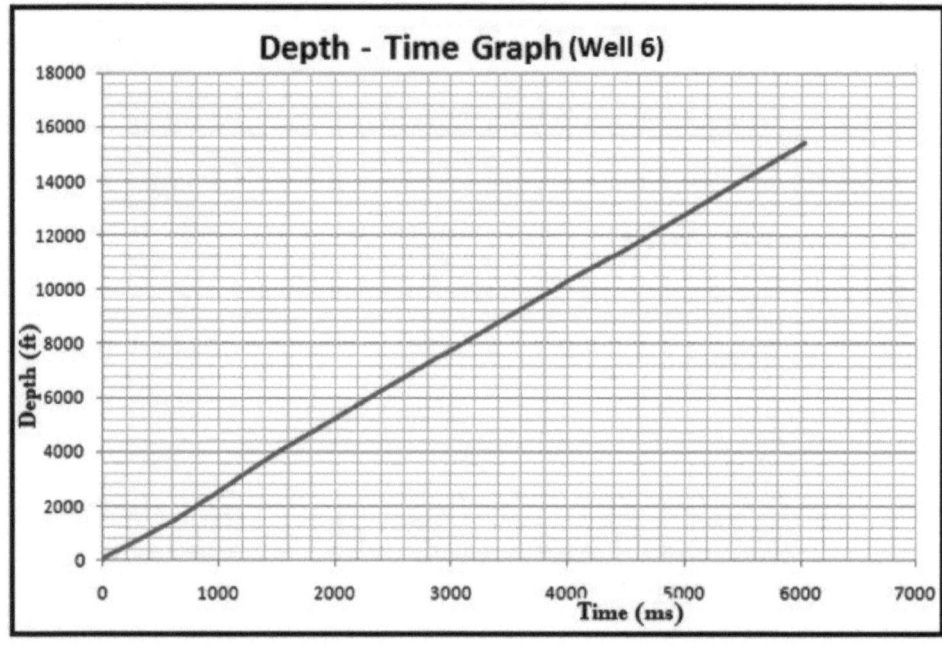

Figure 4.12: The Depth – Time Graph of Well 6.

4.4   WELL LOG INTERPRETATION

The well log Interpretation was done using RocDok Software. Out of the six wells given, four wells were selected for this analysis; these include Tmb 02, 05, 01 and 04. These wells are arranged in this order on the software correlation panel based on their linear occurrence on the base map. Since the area of interest of this research work is thin sandstone beds, Gamma ray and resistivity logs were used exclusively in determining the lithology and the resistivity of the fluid contained within the section of interest (Figure 4. 13). Two thin sandstone beds were successively mapped and their respective tops and base was mapped. Although, the tops and base of the Second thin bed (TB 03T) could not be mapped distinctively.

4.4.1 Thin Sandstone Bed 1 (TB 01T):

The first mapped sandstone bed (TB 01T) was mapped in three of the wells; the fourth well do not seems to contain the sandstone unit (Figure 4.13). This could be due to thinning or tectonic activities. Wells 05, 01 and 04 contains the thin sandstone unit at different depth and thicknesses. Well 2 do not contain the sandstone unit. So, it could not be correlated (Figure 4.14). The sandstone unit is found in well 5 at a nearer depth of 3326.67m (Marker Top) and its base is found at depth 3330.83m with a net thickness of 4.16m (considerably thin). The Gamma ray value at this well shows a negative deflection interpreted as Sandstone- 25.70API, the corresponding resistivity value within this sandstone unit is 44.14 ohm-m. This is relatively high and it is interpreted to contain hydrocarbon.

The sandstone unit in well 5 was correlated with well 1. The sandstone bed was found at the down-dip of well 5 at depth 3364.17m (Marker Top) and the base is found at 3368.75m (Marker

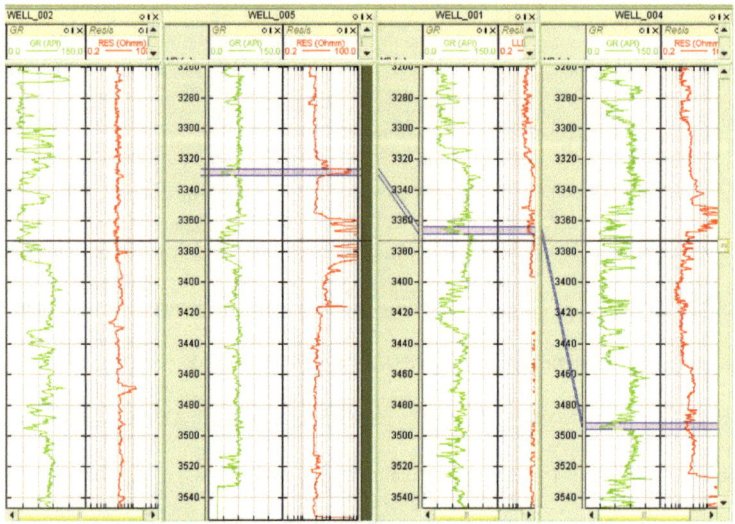

Figure 4.13: Showing the Cross Section of TB 01T across the wells. 7

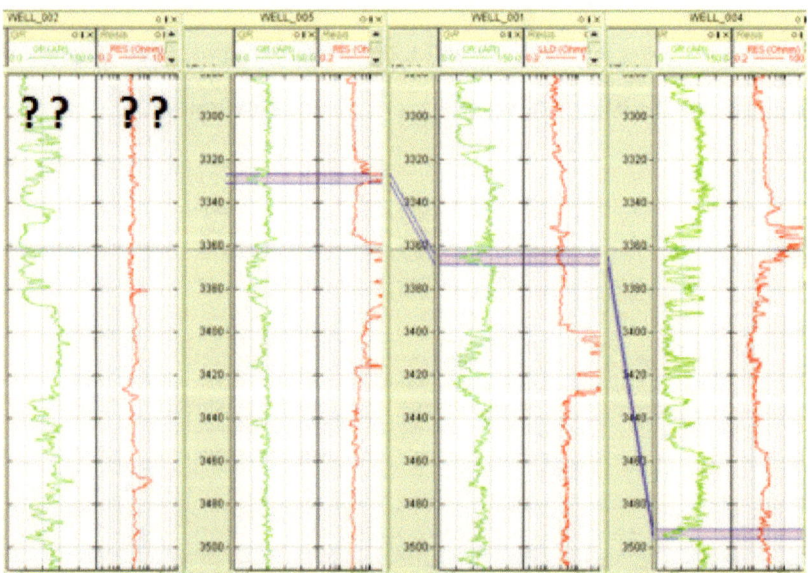

FIGURE 4.14: SHOWING THE CROSS SECTION OF **TB 01T** ACROSS THE WELLS AS IT MISSES OUT IN WELL **02**.

Base). The bed thickness at this depth is calculated to be: 4.58m (considerably thin). The respective gamma ray log value of this thin bed is 39.29API, Interpreted as a sandstone unit, the corresponding resistivity value is 2.15 ohm-m, this is also interpreted to contain hydrocarbon.

At well 4, the same sandstone was also discovered to exist at depth 3491.67m (Marker top) and bounded at the base at depth 3495.83m (Marker base), down-dip of well 01. The bed thickness is estimated to be; 4.16m (Considerably thin also). The gamma ray log deflects to the left with the estimated value of 27.59API, (Interpreted as sandstone). The corresponding resistivity value is 1.85 ohm-m (This is also interpreted to contain hydrocarbon). As earlier mentioned, the traces of thin sandstone 1 could not be found in well 2.

An observation at the thicknesses of this sandstone unit, in relative terms; it could be said that it is of equal thickness (4.16m at Well 5, 5.58m at well 1 and 4.16m at well 4). The devious observation about the thin bed is its depth of occurrence; it varies from one well to the other and in fact completely absent in well 2- it could be suspected that the region has been tectonically influenced. A clear indicator of this is the marginal depth difference in the occurrence depth in well 1 and well 4 (Figure 4.15), (Table 1 shows the tabular presentation of the results and interpretation).

A consideration of the log curve pattern also suggested that the thin sandstone bed is sandwiched between a relatively thick shale unit (which is the normal occurrence in the Niger Delta Basin of Nigeria), (Figure 4.16). The thicknesses of the shale unit varies in the various wells, details are states below;

Table 1 The Properties of Sandstone bed TB 01T in relation to the wells it cuts across

Properties/Wells	Well 5	Well 1	Well 4
Depth of S.S 1- Top	3326.67 m	3364.17 m	3491.67 m
Base	333083 m	3368.75 m	3495.83 m
Thickness	4.16 m	4.58 m	4.16 m
Gamma ray value	25.70 API	39.29 API	27.51 API
Resistivity value	44.14 ohm-m	2.15 ohm-m	1.85 ohm-m
Interpretation	Thin Sandstone unit that contains hydrocarbon	Thin sandstone unit that contains hydrocarbon	Thin sandstone unit that contains hydrocarbon

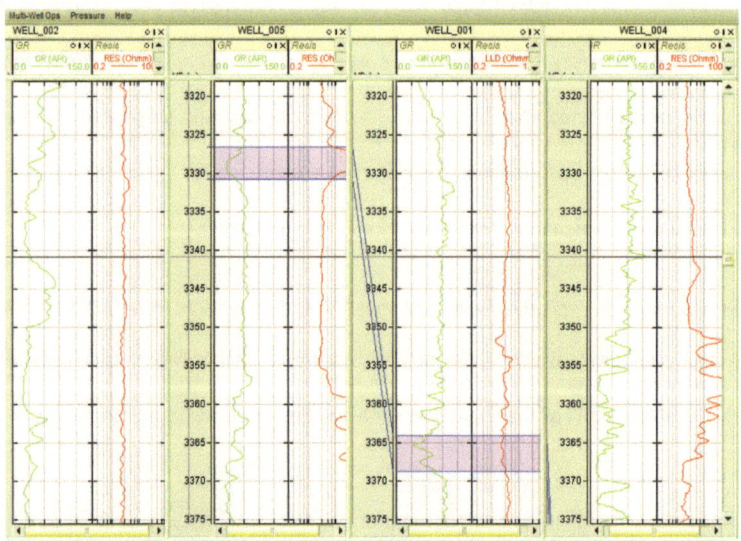

Figure 4.15: Using a Better Resolution Scale to View TB 01T

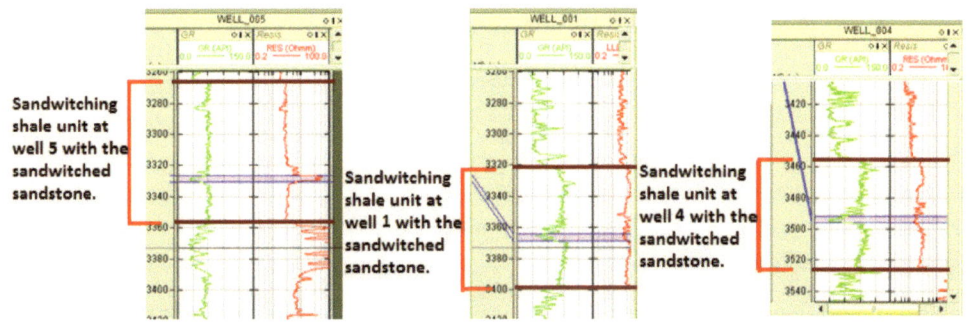

Figure 4.16: TB 01T Sandwiched within a Thick Shale Unit

At well 5:

       Shale top -   3284.38m

       Shale base-   3358.85m

       Shale thickness- 74.85m (Considerably thick)

At well 1:

       Shale top-   3317.71m

       Shale base-   3400.52m

       Shale thickness-   82.81m

At well 4:

       Shale top-   3454.17m

       Shale base-   3527.6m

       Shale thickness-   73.43m

Considering the resistivity values at the respective wells, the sandstone unit could be interpreted to have contained hydrocarbon, the resistivity values are beyond 1 ohm-m and in fact beyond 10 ohm - m in well 5. The fluid type in well 5 could be suggested to be gas because of the relatively high resistivity value, but in this context, it would all be interpreted to contain hydrocarbon.

4.4.2 Thin Sandstone bed 2 (TB 03T):

The second sandstone bed (TB 03T) chosen for this research is extremely thin that its base and top could not be differentiated. An intelligent effort that was made for its mapping was the drastic reduction in the viewing scale which was done at the Z-Scale control panel of the RocDok 4.2.1 software (Figure 4.18). It is somehow interesting that as thin as this sandstone unit is, it was found in all the chosen wells of interpretation unlike the previously mentioned thin bed. Although, it grew thicker and becomes mappable as it approaches well 2 (Figure 4.17).

At well 5, the sandstone bed occurs at depth 2791.15m, Gamma ray value is; 35.73API and the corresponding resistivity value is 1.79 ohm-m. At well 1, the same sandstone unit occurs at depth 2747.92m, gamma ray value is, 42.40API and the corresponding resistivity value is 1.60 ohm-m. At well 2, the sandstone's thickness has increased and it top and base could now be differentiated. The top and base differentiation could not be done because of unison with the other wells. For this reason, the marker line was made to pass through the middle of the bed (Figure 4.18).

Just the same way as it occurred in TB 01T, TB 03T in well 4 occurred at the down-dip of well 1, which confirms the previous suspicion of faulting or tectonic activities in this local region. The gamma ray and resistivity log value are in range as though that of TB 01T. So, the interpretation given to TB 01T above could be related with TB 03T also (Table 2 shows the tabular presentation of the results and interpretation).

TB 03T was also found to be sandwiched between a relatively thick shale unit (Figure 4.19).

Table 2 The Properties of Sandstone bed TB 03T in relation to the wells it cuts across

Properties/Wells	Well 2	Well 5	Well 1	Well 4
Depth of S.S 2-	275052 m	2791.15 m	2747.92 m	2900 m
Gamma ray value	30.89 API	35.73 API	42.40 API	40.63 API
Resistivity value	2.30 ohm-m	1.79 ohm-m	1.60 ohm-m	1.04 ohm-m
Interpretation	Thin Sandstone unit that contains hydrocarbon	Very thin sandstone unit that contains hydrocarbon	Very thin sandstone unit that contains hydrocarbon	Very thin sandstone unit that contains hydrocarbon

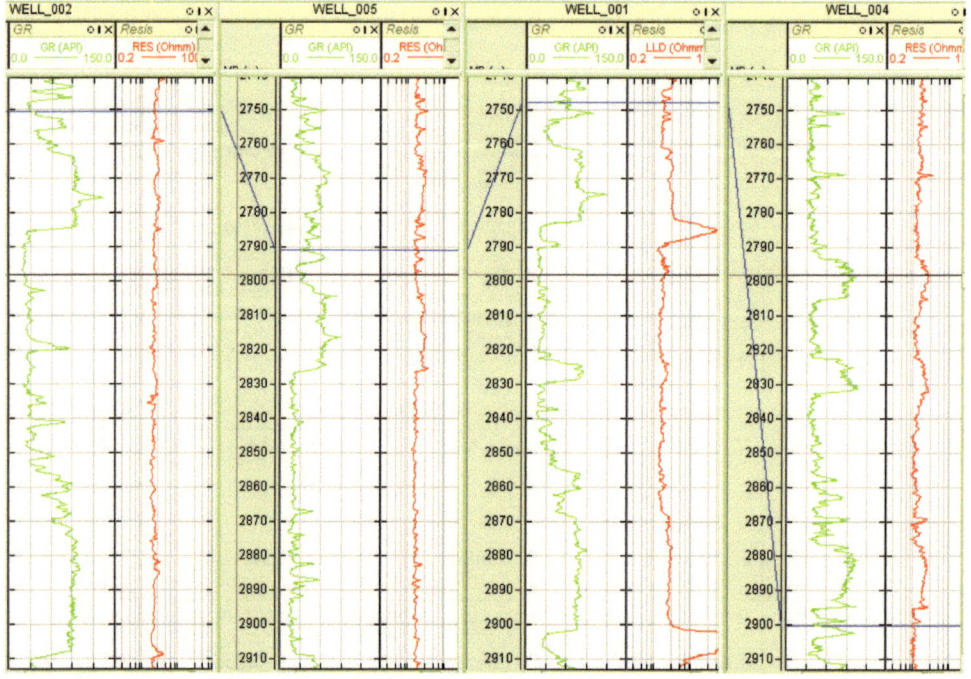

Figure 4.17: Showing the cross section of TB 03T across the wells.

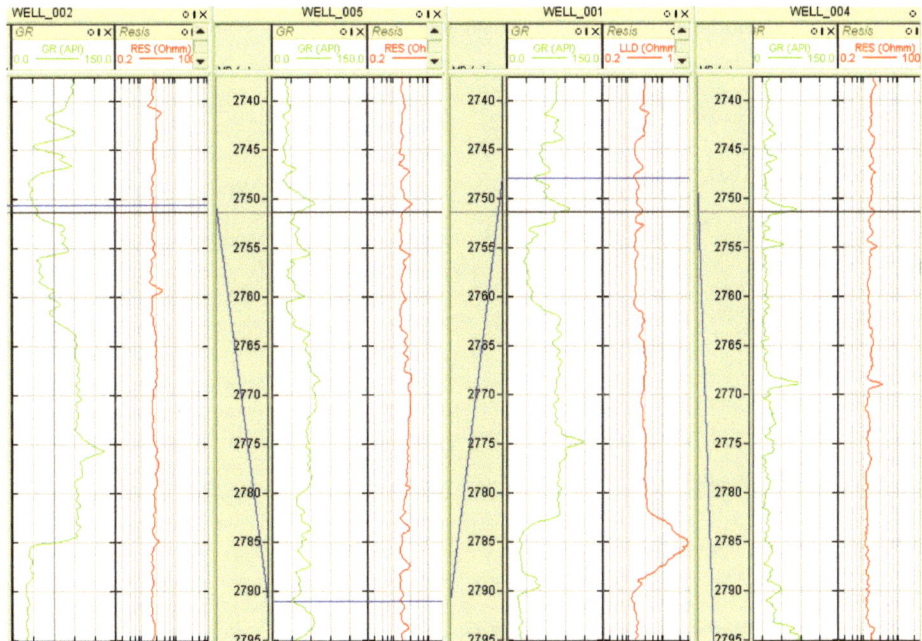

Figure 4.18: An improve scale enhance the viewing and mapping of the TB 03T.

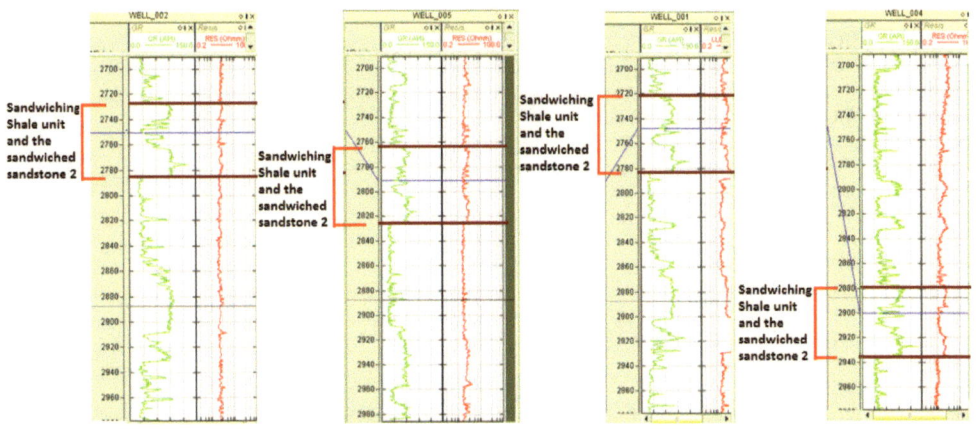

Figure 4.19: TB 03T Sandwiched within a Thick Shale Unit

The estimated details of the sandwiching shale unit are expanded below;

At well 2;

      Shale top: 2727.08m

      Shale base: 2784.9m

      Shale thickness: 52.82m

At well 5:

      Shale top -   2763.02m

      Shale base-   2825.52m

      Shale thickness- 62.5 m (Considerably thick)

At well 1:

      Shale top-   2719.27 m

      Shale base-   2782.29 m

      Shale thickness-   63.02 m

At well 4:

      Shale top-   2876.56 m

      Shale base-   2936.46 m

      Shale thickness -59.9 m

4.5 SPECTRAL DECOMPOSITION

As it has been stated earlier, four wells were used for this analysis (Tmb 02, 05, 01, and 04), Spectral Decomposition (Fast Fourier Transform) was calculated on the seismic sections and the chosen reflectors. The two thin beds picked has been marked on the seismic with a marker name; TB 01T and TB 03T. The TB 03T was found at a shallower depth than TB 01T. Inline 5915 was used exclusively because Well 5 was directly superimposed on it. This makes the interpretation across the volume easier and to reduce errors.

It was noticed on the Seismic sections that the two markers falls on a high amplitude reflector-DHI (i.e. Bright Spot- Figure 4.20). This is a good index for the research and it signals confirmation on the analysis and interpretation that has been done earlier through the well logs. From Well logs, the selected unit is said to be sandstone and has been interpreted to contain hydrocarbon. Further analysis and integration of other interpretation tools has been used and various deductions have been made.

Several attributes were employed and integrated in order to ensure the vivid exposure of the anomaly which is a reference to Geologic features, these attributes include; Instantaneous Amplitude Attribute, Energy Attribute and Spectral Decomposition (Fast Fourier Transform). A color blend was also introduced by combining more than one attribute; all of these are done to enhance the quality of the anomalies.

The Amplitude and Energy attribute (Figures 4.21 and 4.22) shows that the makers fall on a high amplitude black reflector. It was the two markers that fell on this black patched area. This also confirms earlier interpretation done on the well logs. The energy attribute even amplifies the amplitude as it makes it more conspicuous for viewing (Figure 4.22).

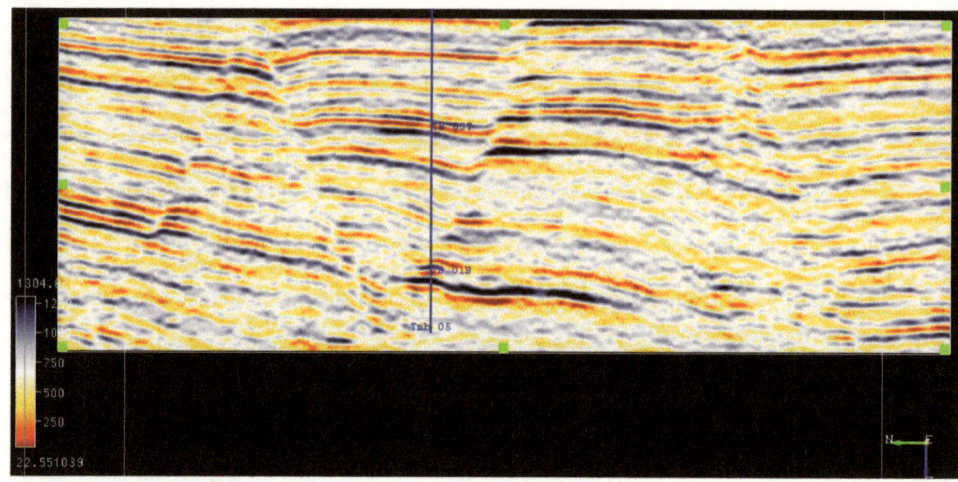

Figure 4.20: Seismic Data Showing the position of the Markers on Inline 5915- Markers falls within the Direct Hydrocarbon Indicator.

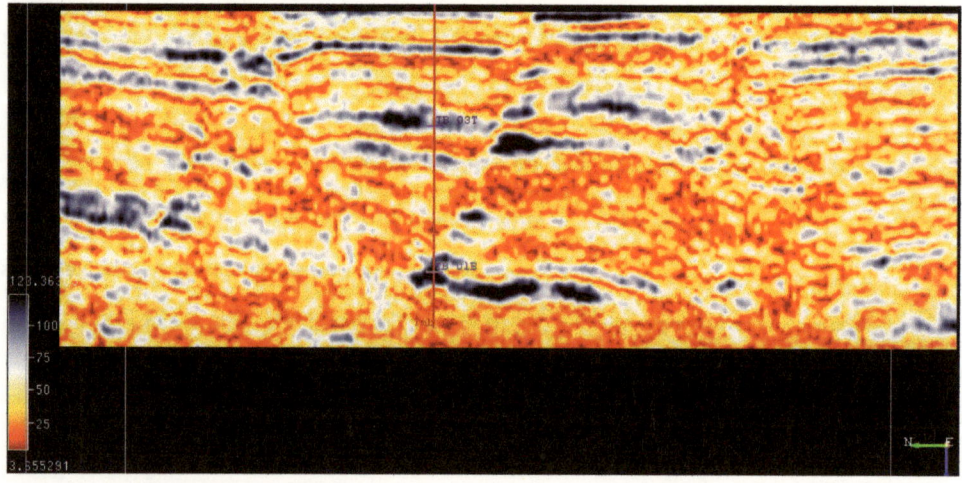

Figure 4.21: Amplitude Attribute- Inline 5915.

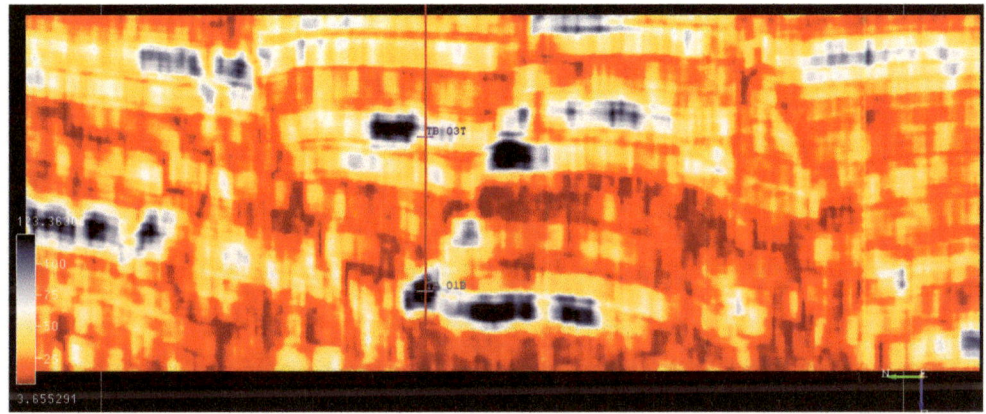

Figure 4.22: Inline 5915- Energy Attribute

4.5.1 Spectral Decomposition Analysis of TB 01T.

Various frequencies were used in mapping the thin bed anomaly, 15Hz, 25Hz, 30Hz, 33Hz, 45Hz and 75Hz (Figures 4.23, 4.24 and 4.25). Prior information about decomposition of frequency has proved that small frequencies are best for mapping thin beds as the objective of this project work is- *"Low frequency components often provide better imaging in poor data areas because the highest signal –to-noise ratio often exists at lower frequencies. Those lower-frequency slices are sometimes able to provide better images of the subsurface"*- Gregory A. Partyka *et. al.*,(Interpretation of 3D Seismic Data- 7th Edition). This was also proved right in this analysis as lower frequencies tend to tune-in the anomaly while higher frequencies tend to tune - out the anomaly, (Figure 4.33 shows the output for 45Hz Fast Fourier Transform, which tends to tune-out the geologic feature).

Spectral Decomposition was calculated on inline 5915 and the results have been presented in the Figures below. The display of 25Hz, 30Hz and 33Hz (Figures 4.23, 4.24 and 4.25) has been presented and these three frequencies seem be the most preferred range where the dominant frequencies exist.

30Hz was chosen to be the most preferred out of the three because the blue patch tends to completely enclose the markers which have been earlier mapped as the region of the thin sandstones (Figure 4.24).

From the spectral decomposition calculations, frequencies beyond 33Hz reveal information other than that of the desired thin sandstone. This was observed as the blue patches gradually disappear on the seismic section (Figure 4.33). However, it was also observed that frequencies below 25Hz tends to depict information other than that of the desired thin.

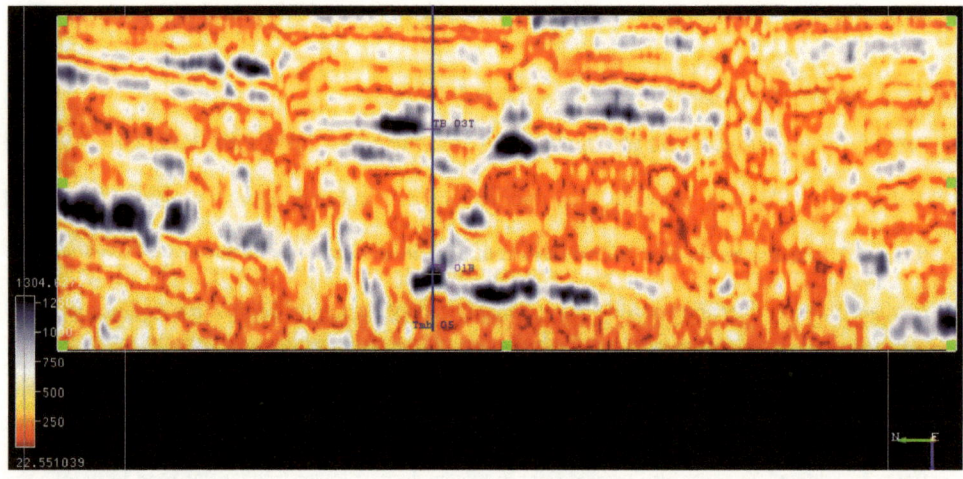

Figure 4.23: Inline 5915- 25Hz FFT

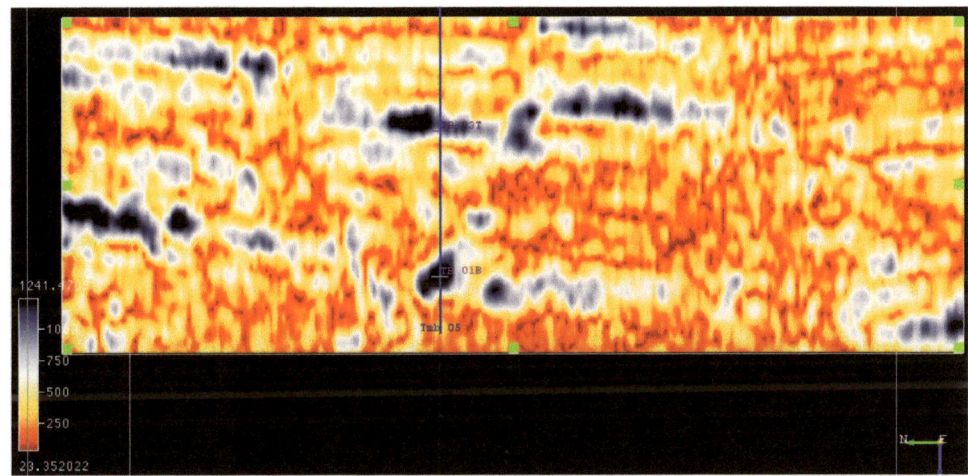

Figure 4.24: Inline 5915- 30Hz FFT

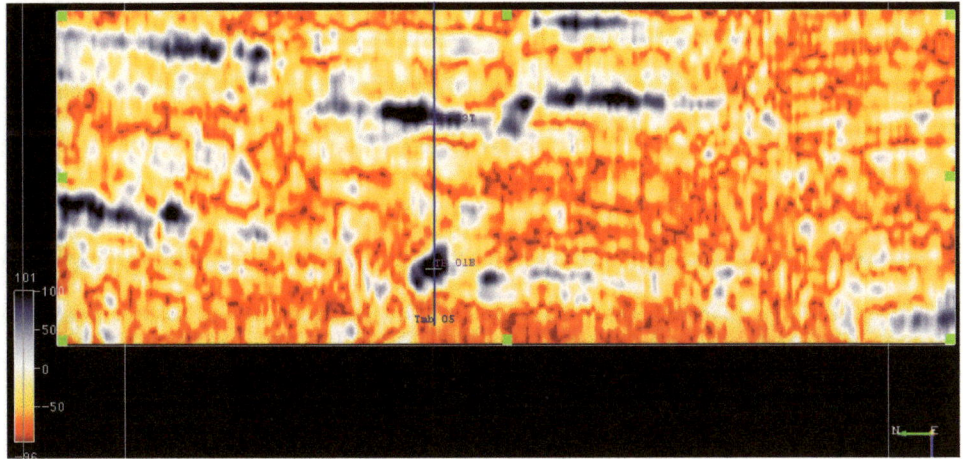

Figure 4.25: Inline 5915- 33Hz FFT

sandstone. It was then resolved that the analysis would concentrate more on frequencies that best map the desired geologic features.

After the conspicuous mapping of the thin bed anomaly with the said frequencies, additional efforts were made by launching the color blend icon of the software in order to combine attributes for display. The selected frequencies that its Spectral Decomposition were calculated are; 25Hz, 30Hz, and 33Hz. These were also used all through the study.

Using the color blend, it enhanced the viewing quality of the anomaly and the bright white patch which depicts the anomaly was distinctively exposed (Figure 4.26, 4.27 and 4.28). Blue was used for 25Hz; Green was used for the 30Hz while Red was used for the 33Hz. However, Energy and Instantaneous attributes were also introduced at this point and it also contributed to the mapping of the anomaly. More importantly, the introduction of these two powerful tools (i. e. Energy and Instantaneous Amplitude) helped confirm the earlier propositions made.

The Z- View of the selected reflector was then made, and the same process was repeated for a lateral view outline of the anomalies. Figures 4.29; shows the Z- View of TB 01T without enhancement in neither frequency nor attribute. Three Wells fell within the blue patched area. These are Tmb 05, 01 and 04. The last well, Tmb 02 falls outside the blue patched area, instead it falls on the orange patched area. This is a strong indication that what these wells would be experiencing at this particular horizon would be quite different. It could be said that the experiences of the first three wells mentioned would be similar while the fourth's experience would be out rightly different from the others. Although, this evidence is not enough to prove that the sandstone only covers the immediate region of the afore mentioned three

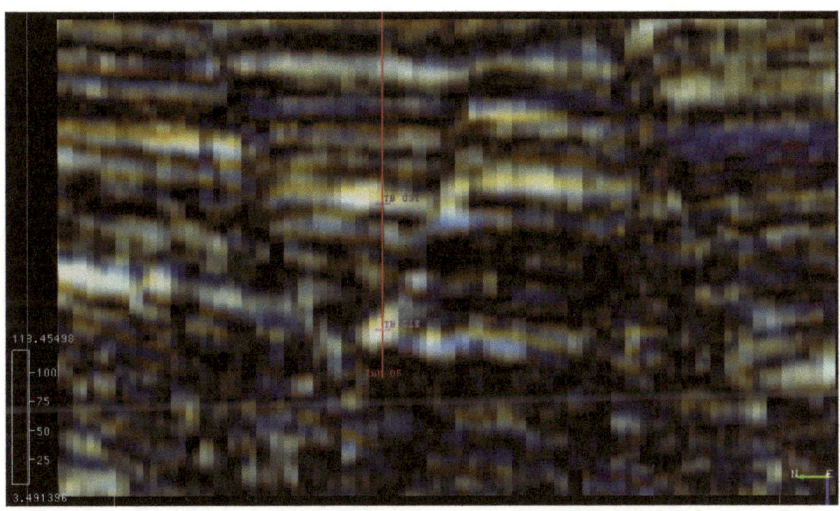

Figure 4.26: Inline 5915- Color blended Image; Blue- 25Hz, Green- 30 Hz, and Red- 33Hz.

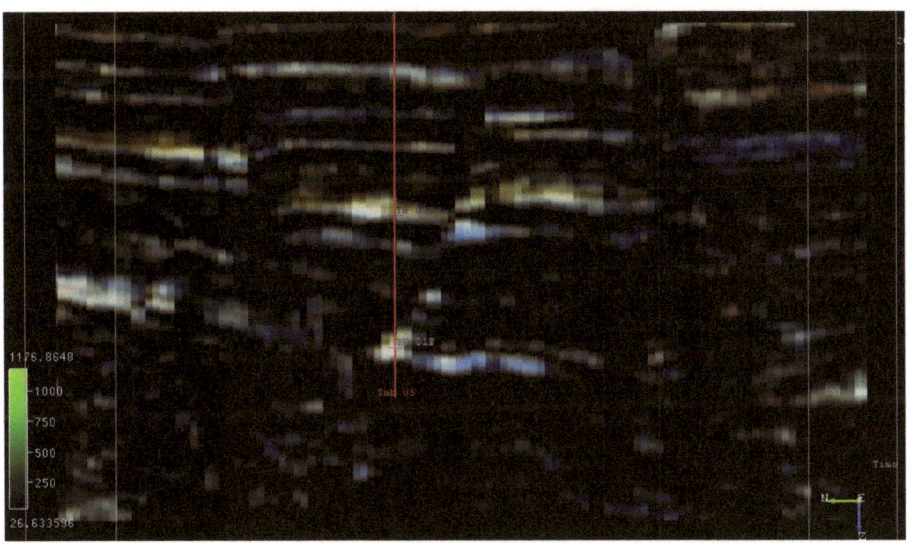

Figure 4.27: Inline 5915- Color blended of Amplitude Attribute, Blue- 25Hz, Green- 30 Hz, and Red- 33Hz.

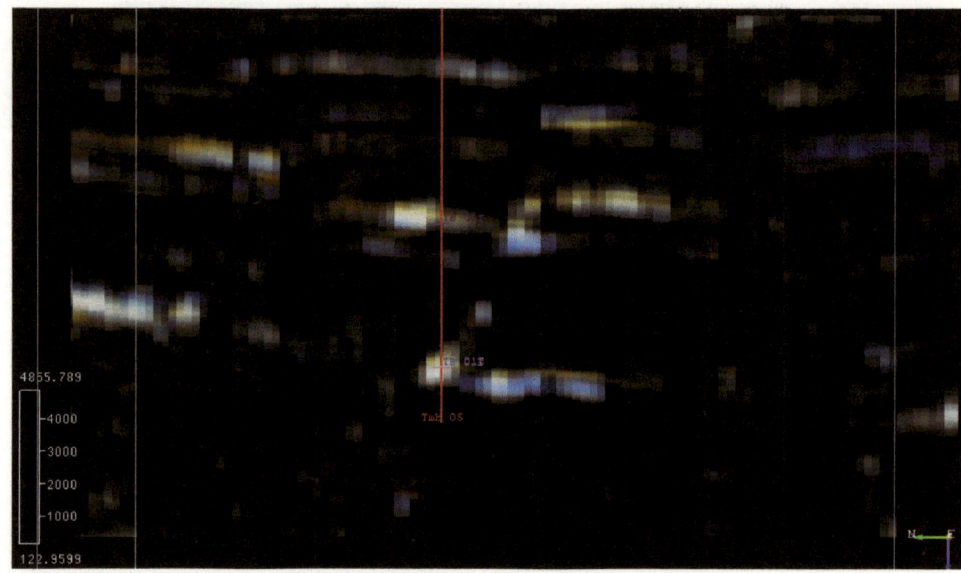

Figure 4.28: Inline 5915- Color blended of Energy Attribute, Blue- 25Hz, Green- 30 Hz, and Red- 33Hz.

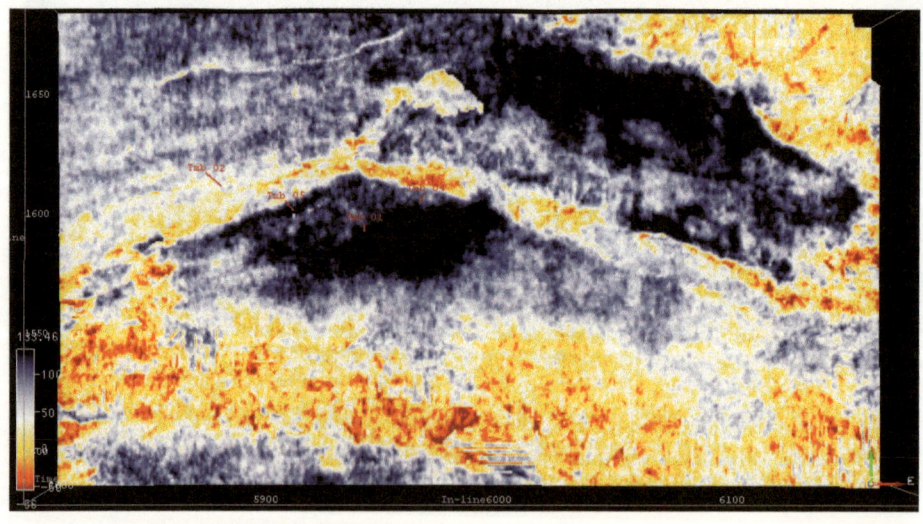

Figure 4.29: The seismic data of TB 01T

wells, but a decomposition of frequency and calculation of other attributes would help more. This has been done and its results are discussed thereafter.

Results from the calculation of 15Hz, 25Hz, 30Hz and 45Hz are shown in Figures 4.30, 4.31, 4.32, and 4.33. 25Hz, 30Hz and 33Hz Frequency results tends to tune in the anomaly more while 45Hz frequency result tune – out the desired features. From observing the results of 25Hz, 30Hz and 33Hz, it was noticed that the earlier mentioned blue patches which serve as our index contain three wells while the fourth well falls outside the blue patch area. This then adds to our first proposition that the experience of the fourth well would be quite different from the experience of the three other wells. A further attempt of mapping the anomaly was then made, Energy and Instantaneous Amplitude attribute was employed and the same scenario repeats itself (Figure 4.34 and 4.35).

This is enough to suggest that the blue patched area is the outline of the thin sandstone. If this is so, a recall on the interpretations from the well logs proved that only three wells contain the sandstone unit and the sandstone unit could not be found in the first well (i. e. TMB 02). This could be suggested to be as a result of thinning, this means during the deposition of this particular sandstone unit, the spread was not extensive and this could have been responsible for its relatively thin net thickness.

The further efforts made was the employment of color blends as it has been employed in the earlier interpretation. Blue was used for 25Hz, Green for 30Hz, Red for 33Hz and Black for either of Energy and Instantaneous Amplitude attribute (Energy and Instantaneous Amplitude were used at different occasions), (Figure 4.36, 4.37 and 4.38).

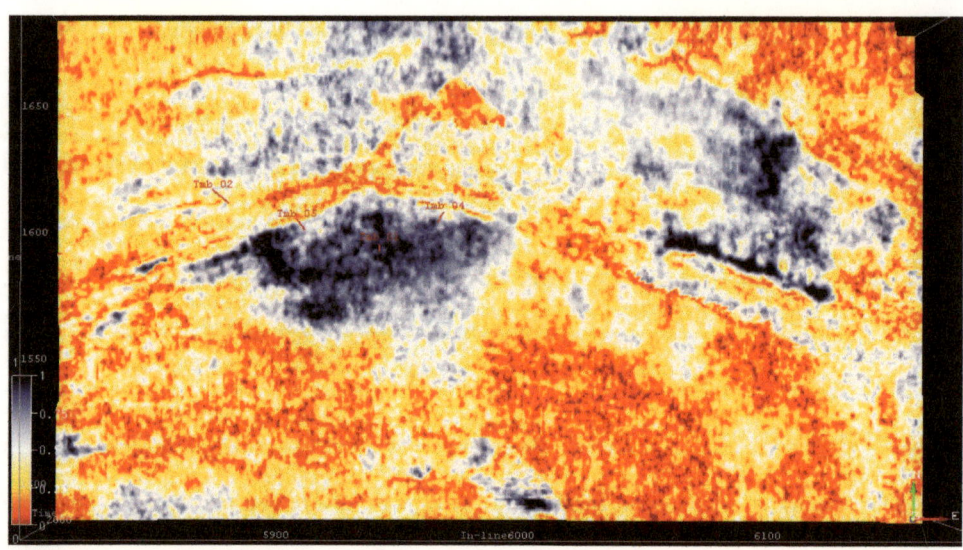

Figure 4.30: The 15Hz FFT of TB 01T

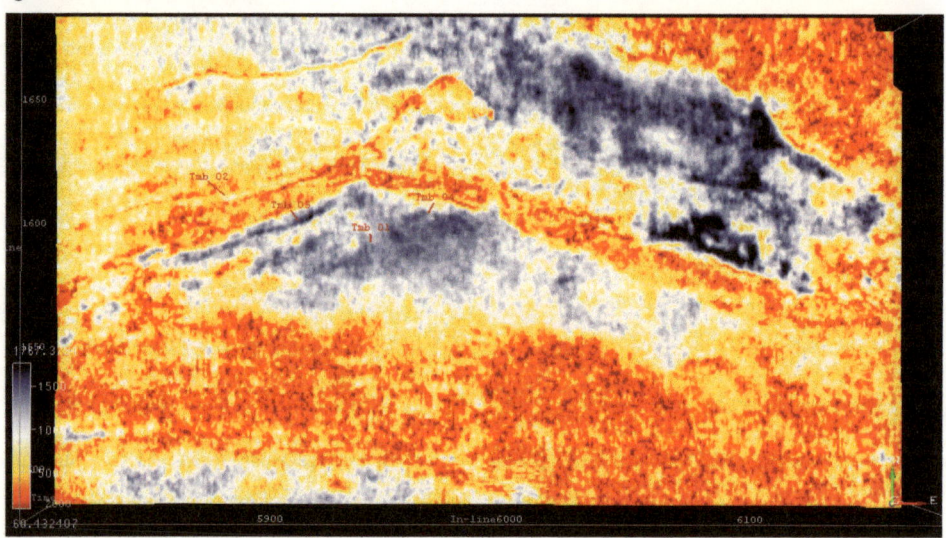

Figure 4.31: The 25Hz FFT of TB 01T

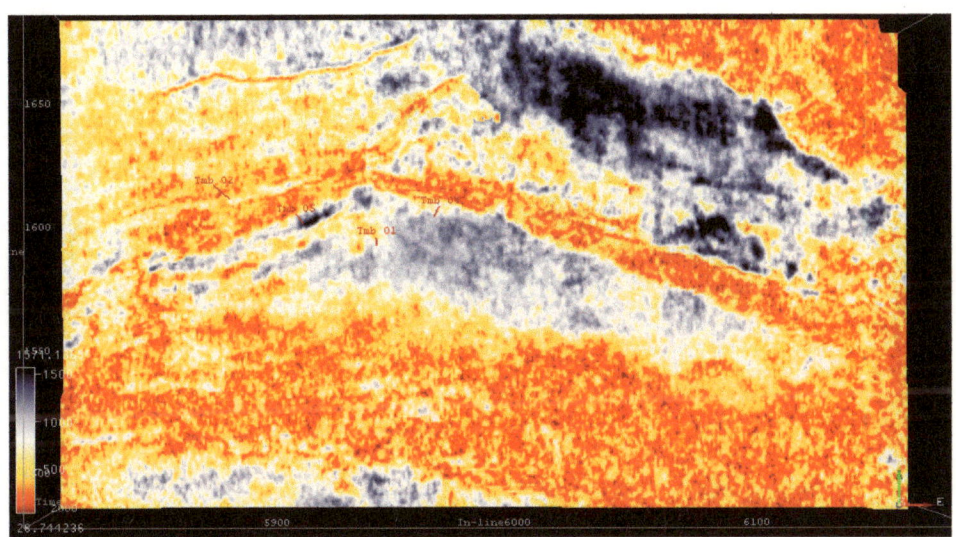

Figure 4.32: Showing the 30Hz FFT of TB 01T

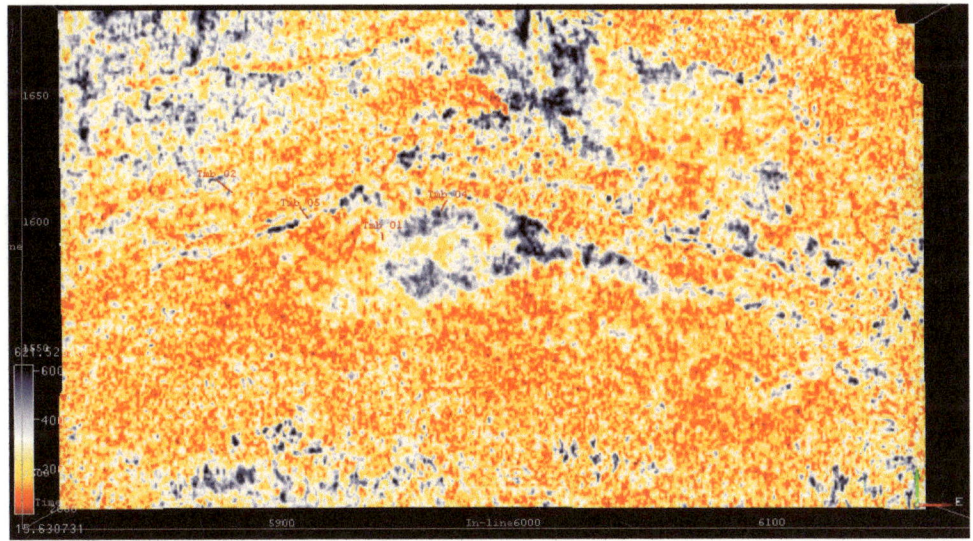

Figure 4.33: Showing the 45Hz FFT of TB 01T

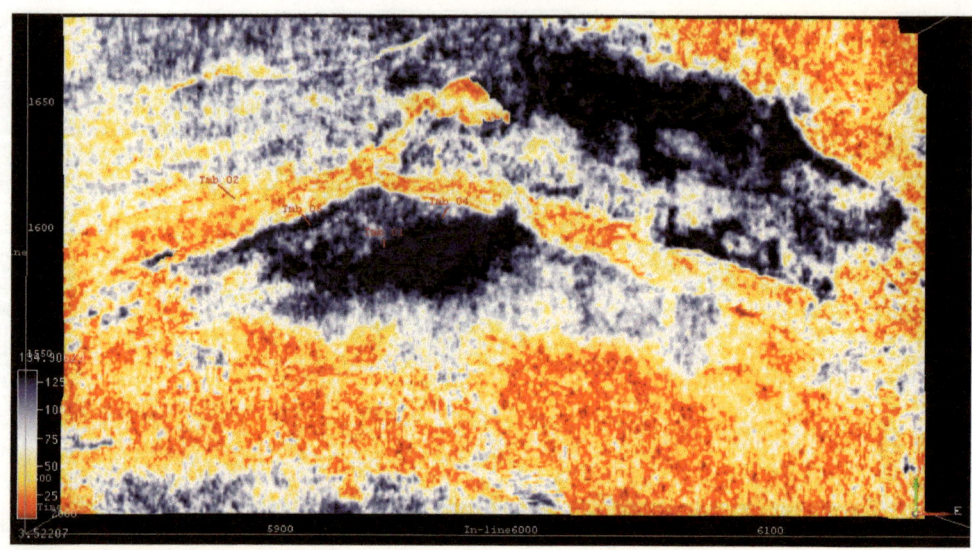
Figure 4.34: Amplitude Attribute Map of TB 01T

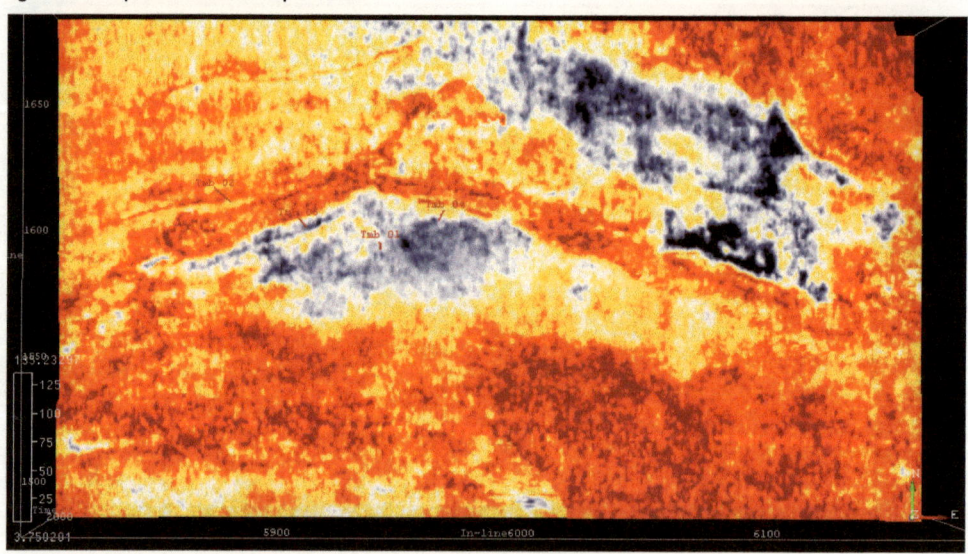
Figure 4.35: Energy Attribute Map of TB 01T

Figure 4.36: Color Blended Map of TB 01T - 25Hz/Blue, 30Hz/Green, 33Hz/Red.

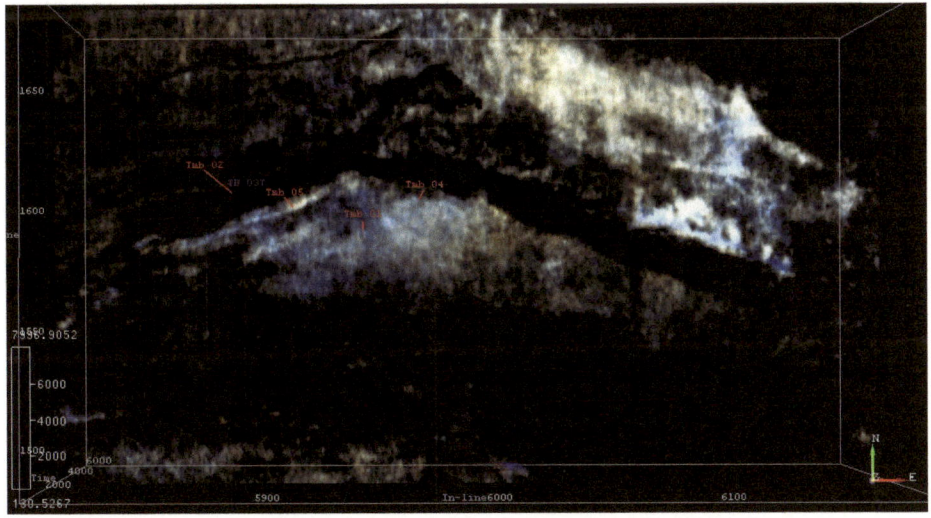

Figure 4.37: Color Blended Map of TB 01T - 25Hz/Blue, 30Hz/Green, 33Hz/Red, with Energy.

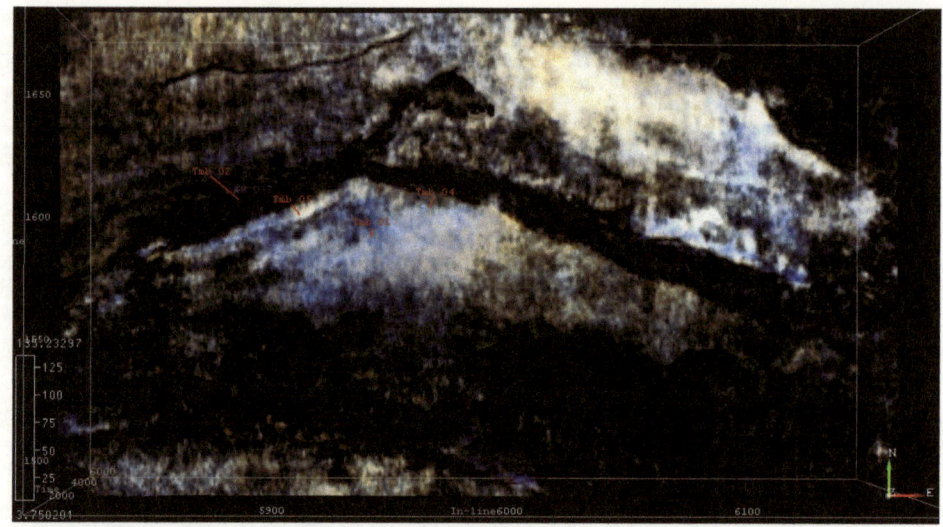

Figure 4.38: Color Blended Map of TB 01T - 25Hz/Blue, 30Hz/Green, 33Hz/Red, with Amplitude.

The physical output for the color blends were quite different from the results that were shown earlier. Some area shows dark - deep brown color, some reflect a bright white with blue stain while some showed a complete black reflection.

In connection with the earlier discussion, the interesting thing is that the region earlier stated through the frequency analysis was the same region of anomaly in the color blend. This further confirms that the sandstone unit is represented by the anomaly. The anomaly is represented on the color blend as a bright white region with a stain of blue. This has been proved that the anomaly region is the outline of the thin sandstone

4.5.2 Trapping Mechanism in TB 01T

An interesting thing about the mapped sandstone unit is the oil trapping mechanism. From the frequency analysis, structures and microstructures has concurrently been mapped. Numerous faults that cuts across this horizon have been distinctively represented in the frequency outputs. From Figure 4.50, it has shown that the hydrocarbon in TB 01T was trapped structurally through the major fault F 1 which cuts across the cube. Suggestively, the faulting has served as the trapping mechanism for the hydrocarbon in TB 01T.

With the above confirmations, new prospects could be identified on TB 01T. On TB 01T, a particular region also has a similar signature as that of the confirmed thin sandstone. This has been noticed on all of the frequency displays. The new prospect region has been named Prospect A. Further confirmation could have been done in this area if a well was drilled and a well log analysis is being carried out.

4.5.3 Spectral Decomposition Analysis of TB 03T.

The same process that was employed in the analysis of TB 01T was also employed in the analysis of TB 03T, although; there were controversies interpreting the outputs. This was due to two major reasons, namely;

1. The sandstone unit is very thin, extremely thin that it could not be distinctively mapped as it was possible for TB 01T.
2. A great stain of shale was suspected in the sandstone unit. This was also proved by the well logs. It has been stated beforehand that the sandstones are sandwithched between a shale units. But the shale that sandwiched TB 03T seems to have more influence on it by its staining effect.

These two factors have contributed to the interpretation that was given to the frequency outputs. Amplitude was first used in detecting the anomaly behavior and their disposition to the wells. Through the instantaneous amplitude attribute (Figure 4.39), it was observed that the TB 03T marker was engulfed by high amplitude crest reflector (i. e. blue patches). This is also an index for the study which requires further proves.

Frequency of 25Hz, 30Hz (Figure 4.40 and 4.41) and 33Hz which has first been considered as the tuning frequencies were also used in analyzing this particular horizon. It was also observed in this particular sandstone bed that its marker falls within the blue patched area. This was also the experience in TB 01T. This simply confirms the previous interpretations that the sandstone unit is represented by the blue patches on the frequency outputs. However, a planar view analysis was also carried out for mapping the geomorphology of the sandstone unit. Instantaneous amplitude attribute was first employed for discrete amplitude mapping (Figure 4.43). Here the blue patches tend to stain nearly all area on the horizon. It was noticed that

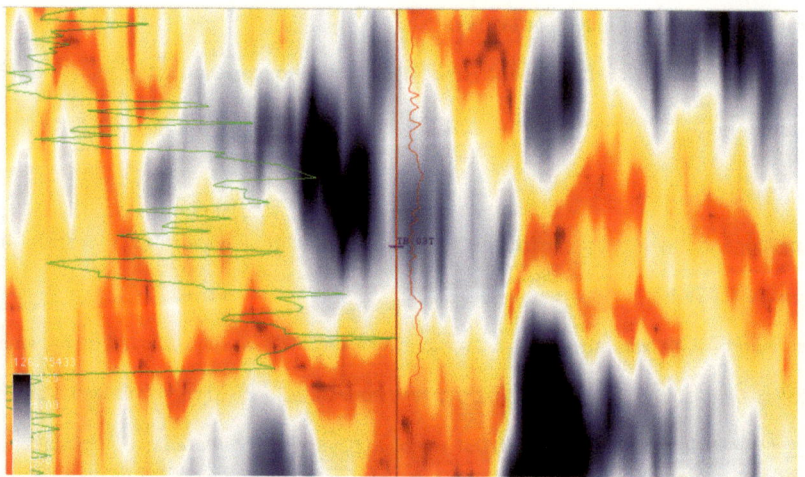

Figure 4.39: Inline 5915 Instantaneous Amplitude Attribute, overlain on it is well 5 and its log and Marker for TB 03T

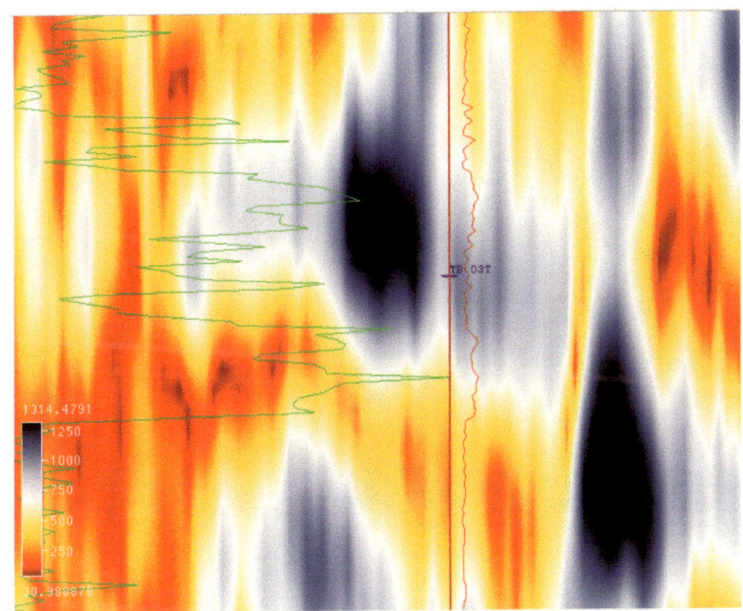

Figure 4.40: Showing Inline 5915 for 25Hz, overlain on it is well 5 and its log and Marker for TB 03T

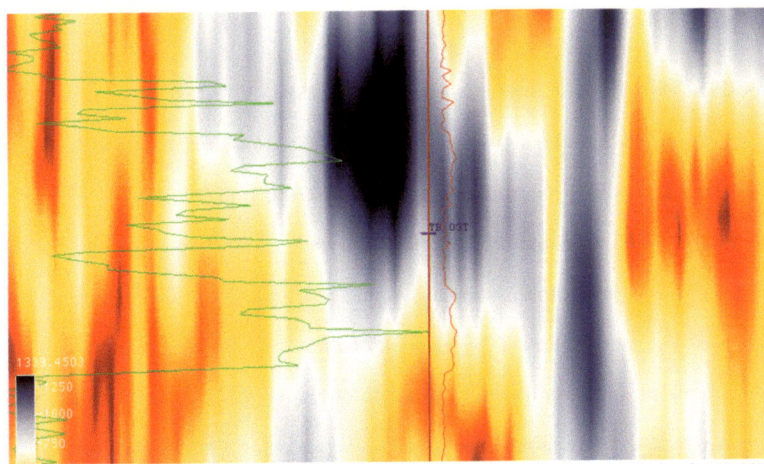

Figure 4 .41: Showing Inline 5915 for 30Hz, overlain on it is well 5 and its log and Marker for TB 03T

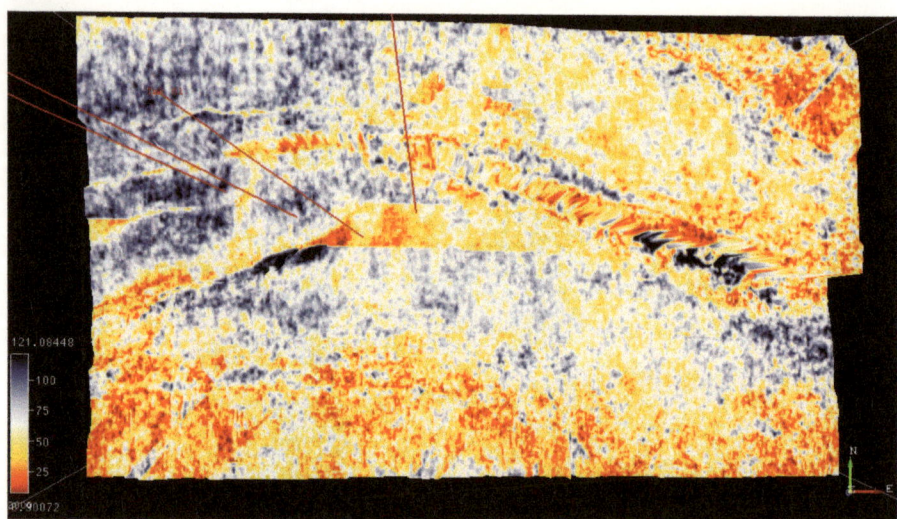

Figure 4.42: Instantaneous Amplitude Attribute on TB 03T.

only two of the wells (Tmb 02, and Tmb 05) fall on the blue patched area while the two other wells (Tmb 01 and Tmb 04) falls on orange patched region. Reasons for this could not be suggested until further studies have been carried out.

Frequency decomposition analysis was carried out on TB 03T. 25Hz, 30Hz, and 33Hz have been applied and its respective outputs have been presented in Figures 4.43, 4.44 and 4.45. It was noticed that Tmb 02 and Tmb 05 fell on a blue patched area while Tmb 01 and Tmb 04 also fall on the blue patched region but it was noticed here that orange color stain were present especially on the 25Hz outputs. A thought in line with the first proposition above that TB 03 has the influence of shale in it could now be introduced. It was evident from the frequency output that the orange color stain that seems to be sparsely sprinkled on the horizon has influence on the whole blue patches that could have been expressed on TB 03T. It was also noticed that the two wells (i. e. Tmb 01 and Tmb 04) are bounded by faults which could have also influenced oil trapping in this horizon.

4.5.4 Hydrocarbon Potential in TB 03T

It has been confirmed in well log analysis that the TB 03T contains hydrocarbon, spectral analysis has also showed the same anomaly signature that was recognized for the thin sandstone and the presence of shale has been proved. The hydrocarbon contained in this sandstone unit would be chiefly under these two influence;

   a. Thickness and b. Presence of shale

The thickness of this sandstone unit has been shown in well log analysis to be very thin and even seems un-mappable, this might affect the quantity of oil contained by the sandstone unit. Also, presence of shale in reservoirs has been proved to always reduce its Hydrocarbon potentials due to the porosity and permeability reduction effect that it produces.

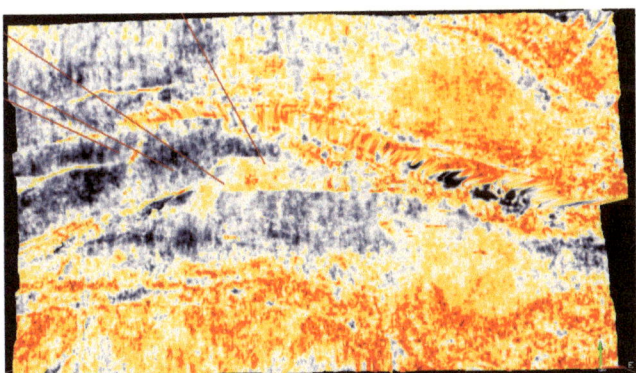

Figure 4.43: 25Hz Frequency display on TB 03T

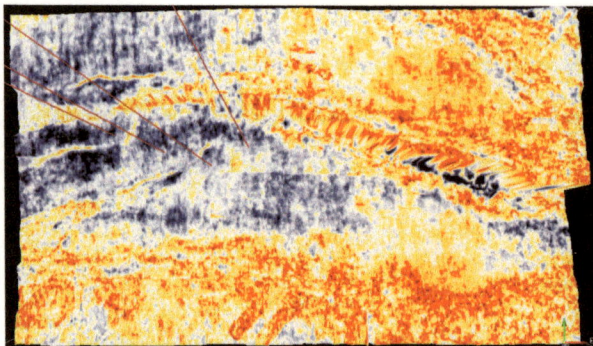

Figure 4.44: 30Hz Frequency display on TB 03T

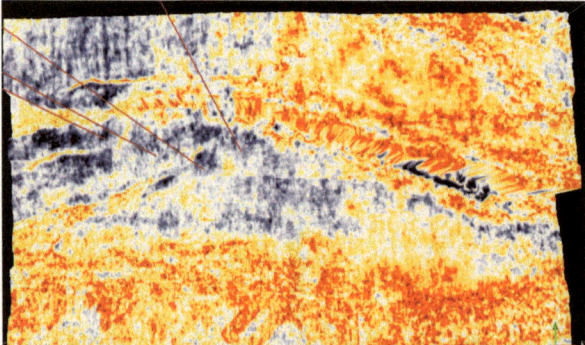

Figure 4.45: 33Hz Frequency display on TB 03T

4.6 ISO - CHRON MAPS

After the Spectral analysis has been done, the time values for each horizon were transferred to the base map using commands on the OpendTect software. It should be noted that one of the limitations that was encountered in this study was that the un – licensed version of OpendTect software was used in this study, this limited the access to some plugins. The plugin which is meant for producing map (GMAT) could not be used but the effort made was to extract the data from the software and export to Oasis Montag. The map production and contouring was done using Oasis Montag. So, the faults and well locations could not feature on the produced map.

The isochron maps which shows the variation in time across the field (Figures 4.46 and 4.47) reveal the presence of roll-over structures in the area covered by the sections. Weber and Daukoru (1975) discussed that the presence of roll-over structures may indicate possible hydrocarbon accumulation in the vicinity of such roll-over structures. Therefore, the presence of faults and roll-over structures in the study area suggest possibility of hydrocarbon accumulation. This further proves the earlier proposition that Hydrocarbon is present in the picked region and even the suspected prospect. A time closure was observed in the Prospect A region. (Figure 4.46)

4.6.1 TB 01T ISOCHRON MAP: Time values to this horizon range 2437.4 ms to 2892.1 ms

Time-contour closures can be seen on the horizon. These closures indicate roll-over structures. It was also noticed that the area picked as prospect A also has a time closure.

4.6.2 TB 03T ISOCHRON MAP: Time values to this horizon range 2163.3 ms to 2457.3 ms.

Time-contour closures (structural closures) can also be seen on this horizon, although fault controlled closures. The structural closures also indicate roll-over anticline.

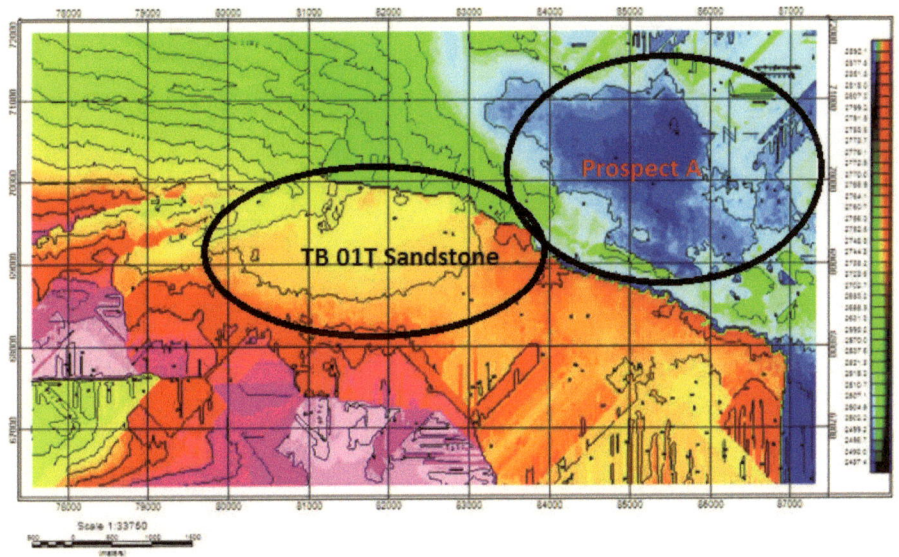

Figure 4.46: Iso – Chron Map for TB 01T

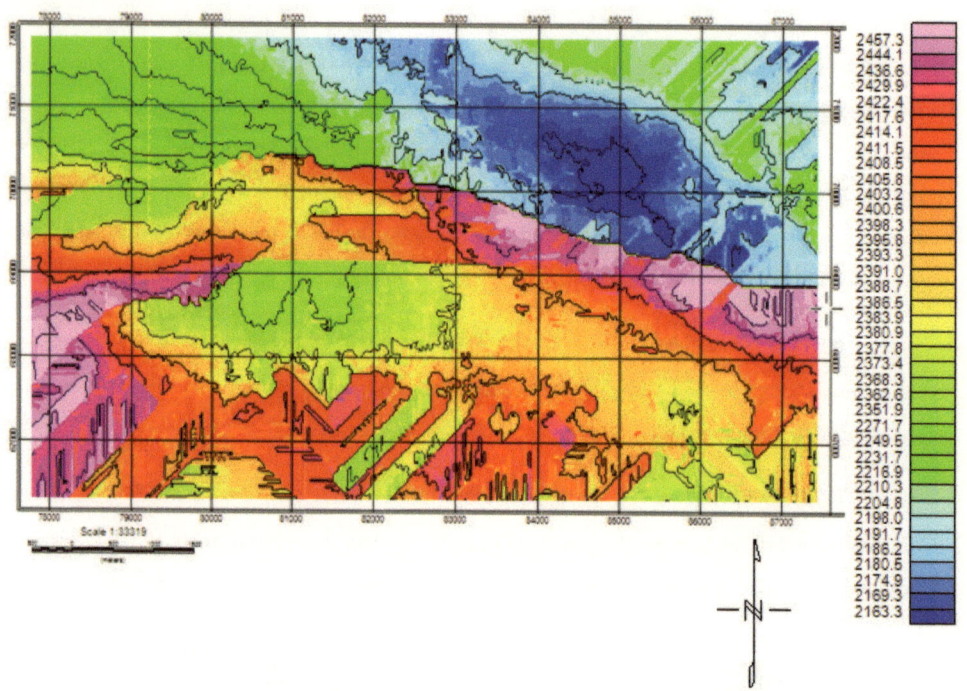

Figure 4.47: Iso-Chron Map for TB 03T

CHAPTER 5

5.1 CONCLUSION

The following are the deductions made from the study:

The conventional Interpretation has helped to map the existing faults and the pay sand reflectors in the seismic volume. Well log analysis has also helped in mapping the hydrocarbon contained sandstone and their relation to depth.

Spectral Decomposition has helped in mapping pay sandstone sectionally. However, the plan view analysis has helped map the thin pay, its geomophology, facies distribution and the wells disposition to the pay sands. Thin Sandstone reservoir facies and non reservoir facies have been differentiated by the integration of well logs and modern 3D interpretation tools. Spectral Decomposition has also helped in mapping new prospects. Although, nearly all the Findings has been qualitatively validated.

It has also been validated through the integration of well log analysis and spectral decomposition analysis that TB 01T is more prospective that TB 03T. This is responsible for the earlier stated reasons that TB 03T has more of shale stain and is extremely thin that its contained hydrocarbon might not be of commercial quantity. The Iso – Chron maps has also shown time closure in the mapped thin sandstone region which indicate roll over structures which could be an hydrocarbon trap as earlier discussed by Weber and Daukoru (1975).

5.2 RECCOMENDATION

As a result of the succinct work carried out on the chosen thin sandstone units, the following recommendations could be made;

a. The suggested prospect be further explored, this could through the drilling of an exploratory well.
b. Quantitative analysis could be carried out on the delineated beds, this would even further confirm the deductions from the spectral decomposition analysis.

REFERENCES

Avbovbo, A.A. (1978) Tertiary lithostratigraphy of Niger Delta: AAPG Bulletin, v.62, p. 295-300.

Beka, F. T., and Oti, M. N. (1995) The distal offshore Niger Delta: frontier prospects of a mature petroleum province, *in*, Oti, M.N., and Postma, G., eds., Geology of Deltas: Rotterdam, A.A. Balkema, p. 237-241.

Blumentritt, C. H. (2008) Highlight volumes: Reducing the burden of interpreting spectral decomposition data: The Leading Edge, **27**, 330–333.

Burke, K. (1972) Longshore drift, Submarine canyons and Submarine fans in development of Niger delta: American Association of Petroleum Geologists Bulletin, v. 56, p. 1975-1983.

Castagna, J. P., S. Sun, and R. W. Siegfried (2003) Instantaneous spectral analysis: Detection of low-frequency shadows associated with hydrocarbons: The Leading Edge, **22**, 120–127.

Doust, H. (1989) the Niger delta: Hydrocarbon potential of a major Tertiery province: proceedings, KNGMG symposium "Coastal Lowlands, Geology and Geotechnology." 1987: Dordrecht, Kluwer, p. 203-212

Doust, H. and Omatsola, E. (1990) Niger Delta, in Edwards, J. D. And Santogrossi, P.A., Divergent/Passive margin basins. AAPG memoir, 48, p. 239-248.

Ejadwwe, J. E. et al. (1984). Evolution of oil generative window and an oil and gas occurrence in Tertiary Niger Delta basin: AAPG Bulletin, v.68, p. 1744 – 1751.

Ekweozor, C. M., and Daukoru, E. M. (1984) Petroleum source bed evaluation of Tertiary Niger Delta--reply: American Association of Petroleum Geologists Bulletin, v. 68, p. 390-394.

Ekweozor, C. M., and Daukoru, E.M. (1994) Northern delta depobelt portion of the Akata-Agbada petroleum system, Niger Delta, Nigeria, *in*, Magoon, L.B., and Dow, W.G., eds., The Petroleum System—From Source to Trap, AAPG Memoir 60: Tulsa, American Association of Petroleum Geologists, p. 599-614.

Ekweozor, C. M., Okogun, J.I., Ekong, D.E.U., and Maxwell J.R. (1979) Preliminary organic geochemical studies of samples from the Niger Delta, Nigeria: Part 1, analysis of crude oils for triterpanes: Chemical Geology, v 27, p. 11-28.

Ekweozor, C.M., and Okoye, N.V. (1980) Petroleum source-bed evaluation of Tertiary Niger Delta: American Association of Petroleum Geologists Bulletin, v. 64, p 1251-1259.

Hospers, J. (1965) Gravity field and structure of the Niger Delta, Nigeria, West Africa: Geological Society of American Bulletin, v. 76, p. 407-422.

Ekweozor, C. M. and Dankoru, E. M. (1980) Petroleum source bed evaluation of Tertiary Niger Delta: AAPG Bulletin, v.68, p. 390-394.

Ekweozor, C.M. and N.V. Okoye, (1980) Petroleum source bed evaluation of Tertiery Niger delta: AAPG Bulletin, v. 64, p. 1251-1259.

Evamy, B.D. et al. (1978). Hydrocarbon habitat of tertiary Niger Delta: AAPG Bulletin, v. 62 p. 1-39.

Harilal, C. G. Rao, R. C. P. Saxena, J. L. Nangia and N. K. Verma, 2004) 3D Seismic Delineation of Thin Sandstone Reservoirs in Shale-Limestone Rich Sequence of Tapti-Daman Area: A modeling Aided Approach: 5th Conference & Exposition on Petroleum Geophysics, Hyderabad-2004, India P. 423-430

Partyka et. al., (2010) AAPG Memoir 42 SEG Investigations in Geophysics, No. 9 Interpretation of Three-Dimensional Seismic Data Seventh Edition.

John P. Castagna, **Shengjie** Sun and **Robert** W. Siegfried, 2002, the Use of Spectral Decomposition as a Hydrocarbon Indicator: Gas TIPS, PP 25-28.

Kaplan, A., Lusser, C.U., Norton, I.O. (1994) Tectonic map of the world, panel 10: Tulsa, American Association of Petroleum Geologists, scale 1:10,000,000.

Klett, T.R., Ahlbrandt, T.S., Schmoker, J.W., and Dolton, J. L. (1997) Ranking of the world's oil and gas provinces by known petroleum volumes: U. S. Geological Survey Open-file Report-97-463, CD-ROM.

Knox, G.J., and E.M Omatsola 1989, Development of the Cenozoic Niger delta in terms of the "Escalator Regression" model, and impact on hydrocarbon distribution: proceedings KNGMG symposium "Coastal Lowlands, Geology and Geotechnology," 1987: Dordrecht, Kluwer, p. 181-202.

Kulke, H. (1995) Nigeria, *in,* Kulke, H., ed., Regional Petroleum Geology of the World. Part II: Africa, America, Australia and Antarctica: Berlin, Gebrüder Borntraeger, p. 143-172.

Lambert-Aikhionbare, D. O., and Ibe, A.C. (1984) Petroleum source-bed evaluation of the Tertiary Niger Delta: discussion: American Association of Petroleum Geologists Bulletin, v. 68, p. 387-394.

Lehner, P., and De Ruiter, P.A.C. (1977) Structural history of Atlantic Margin of Africa: American Association of Petroleum Geologists Bulletin, v. 61, p. 961-981.

Marfurt, K. J., and **Kirlin**, R. L., 2001, Narrow-band spectral analysis and thin-bed tuning, Geophysics, no. 4, 1274 –1283.

Matt Hall and ***Eric*** Trouillot, 2004, Predicting stratigraphy with spectral decomposition: 2004 CSEG National Convection.

Merki, P.J., Structural geology of the Cenozoic Niger delta, in 1st conference African proceedings: Ibadan university press, p.251-266.

Meyer, D. et al., (2000). Use of seismic attributes in 3-D geovolume interpretation. *The Leading Edge*, December 2001.

Mohamed A. Eissa, John P. Feiffer and Heiman Alfredo Paz Ortega, 2009, Seismic Petrophysical Analysis for Thin Sandstone Reservoirs in Columbia's Guajira Basin: TLE, 2009, PP 640 – 647

Nwachukwu, S.O., and P.I Chukwura, 1986, Organic matter of Agbada Formation, Niger delta, Nigeria: AAPG Bulletin, v. 70, p.48-55.

Orife, T.M., and A.A. Avbovbo, 1981, Stratigraphic and unconformity traps in the Niger delta (abs.): AAPG Bulletin, v. 65, p. 967.

Partyka, G. A., J. A. Gridley, and J. A. Lopez, (1999) Interpretational aspects of spectral decomposition in reservoir characterization: The Leading Edge, **18**, 353–360.

Petroconsultants, (1996) Petroleum exploration and production database: Houston, Texas, Petroconsultants, Inc., [database available from Petroconsultants, Inc., P.O. Box 740619, Houston, TX 77274-0619].

Peyton, L., Bottjer, R., and Partyka, G. (1998) Interpretation of incised valleys using new 3-D seismic techniques: A case history using spectral decomposition and coherency: *The Leading Edge,* no. 9, 1294-1298.

Puryear, C. I., and J. P. Castagna, (2008) Layer thickness determination and stratigraphic interpretation using spectral inversion: Theory and application: Geophysics, **73**, R37–R48.

Reijers, T.J.A *et al.* (1997). The Niger Delta Basin, in Selley, R.C, ed, African Basin-sedimentary Basin of the world 3: Amsterdam, Elsevier Science, p. 151 – 172.

Short, K.C and Stauble, A.J. (1967). Outline of geology of Niger Delta: AAPG Bulletin, v.51, p. 761 – 779.

Sinha, S., P. S. Routh, P. D. Anno, and J. P. Castagna, (2005) Spectral decomposition of seismic data with continuous-wavelet transforms: Geophysics, **70**, 19–25.

Stacher, P. (1995) Present understanding of the Niger Delta hydrocarbon habitat, *in* , Oti, M.N., and Postma, G., eds., Geology of Deltas: Rotterdam, A.A. Balkema, p. 257-267.

Van Hoek, T., and B. Salomons, (2006) Understanding the seismic expression of complex turbidite reservoirs through synthetic seismic forward modeling: 1D convolutional versus 3D modeling approaches: 26th Annual Gulf Coast Section SEPM Foundation Bob F. Perkins Research Conference, 345–372.

Weber, K.J., and E. Daukoru, (1975) Petroleum geology of the Niger delta: Tokyo, 9th world petroleum congress proceedings, v.2, p. 209-221.

Whiteman, (1982) Petroleum geology of Nigeria: Journal of African Earth Sciences, vol. 1, issue 2, pp. 177-180.

X. D. Wei, (2010) Interpretational Applications of Spectral Decomposition in Identifying Minor Faults: 72nd EAGE Conference & Exhibition, Barcelona, Spain, 14 - 17 June 2010.

Xiao, H., and Suppe, J. (1992) Origin of rollover: American Association of Petroleum Geologists Bulletin, v. 76, p. 509-22.